Contents

Preface

Even if there were no other reason than the ever-increasing price of all kinds of energy, the days when one could waste this precious commodity are well past. Fossil fuels are a wasting asset. They are stored solar energy which beamed down upon the earth millions of years ago, and within not too many decades, there will be very little left.

Eventually widespread and finally exclusive use of nuclear power will solve our energy problems—but not yet. Meanwhile we are faced by ever-increasing prime energy costs.

This book shows how we can use modern technological advances to save money and preserve our fossil fuel stocks. The methods given are not the pie-in-the-sky utopian ideas so beloved by arts-trained idealists; they are sound and well-proved engineering solutions which have been chosen because they are economically and financially viable.

I wish to thank the numerous industrial firms who have helped me by supplying information and allowing me to use illustrations. I wish them well in their marketing endeavours. Their enterprise will help us to bridge the gap between the exhaustion of fossil fuels and the ultimate development of cheap and reliable nuclear power.

University of Salford 1983
R. M. E. Diamant

1 The overall strategy of energy saving

It is impossible to forecast future economic trends accurately except in one respect: if a non-renewable asset is being used up, its price is bound to go up relative to everything else.

The vast bulk of the energy used in the world today is in the form of non-renewable oil, natural gas and coal. These resources were laid down many millions of years ago and are at present being consumed at the rate of almost 9 billion tonnes of coal equivalent or 0.32 million petajoules (1 petajoule = 10^{15} joules) per annum, **Table 1.1**.

This compares with proved recoverable world resources of 20.3 million petajoules of coal and 11.2 million petajoules of oil and natural gas. However, this is not the whole story. Advanced capitalist industrial countries, which make up to 15.75 per cent of the world's population, consume 54.5 per cent of these 0.32 million petajoules each year. It is unlikely that the world's underdeveloped countries are going to achieve the living standards of the opulent west overnight. If they did, the annual world's energy demand would rise to 1.11 million petajoules each year even if the living standards in all western countries remained constant. Obviously, looked at from the point of view of western society, things are not particularly desperate as yet. There may be more coal and oil to be found than we know of at present. The snail's pace development of fast-breeder nuclear reactors and fusion reactors may yet be accelerated. Also the poor, with their very modest demands for the earth's stored energy resources, will be with us for many years yet.

On the other hand, although actual energy shortages are not likely to be experienced for many decades, it is certain that since fossil fuels are a wasting asset (like land, coal and oil are not being made any longer), costs are bound to go up. For this reason we must abandon many wasteful practices in energy utilization which have grown up over the years, particularly during periods when fossil fuel supplies seemed to be inexhaustible.

We must conserve energy by all possible means as a bounden duty to those who come after us, our as yet unborn descendants. We must also strive after good energy economy out of quite hard-headed and commercial considerations, to reduce manufacturing costs and stay competitive vis-a-vis our commercial rivals. In 1976 British industry used 2410 petajoules (10^{15} J) of energy: that is, 66.95 million tonnes of coal equivalent. Potential energy savings are as in **Table 1.2**.

TABLE 1.1 WORLD ENERGY USE IN 1978

	Population in millions	Percentage of total world population	Kg of coal equivalent per head	Total in million tons of coal equivalent	Energy used as percentage of total
Underdeveloped countries	1 300	30.5	161	209.3	2.4
Middle income countries	873	20.5	903	788.3	9.1
Advanced capitalist industrial countries	668	15.75	7060	4716.1	54.5
Authoritarian communist countries	1 353	31.85	2177	2945.5	32.9
Oil producing countries without major industries	60	1.4	1620	97.2	1.1
Total:	4 254	100.00		8756.4	100.0

Source: A. McKillop 'Global economic change and new energy' *Energy Policy* 9,4 p. 229–235 Dec. 1981

TABLE 1.2 POTENTIAL SCOPE FOR ENERGY SAVINGS IN BRITISH INDUSTRY

Type of energy saving	Million tons coal equivalent/ annum	Petajoules $(10^{15}$ J) per annum
Waste heat recovery	6 – 8	216–288
Waste as fuel	3 – 5	108–180
Improved instrumentation and control	2 – 3	72–108
Heat pumps	1	36
Process insulation	1	36
Improved drying and evaporation practice	1	36
Industrial combined heat and power plants	1	36
Improved methods of driving machinery	2 – 3	72–108
Total savings:	17–23	612-828

Source: a committee appointed by the Energy Technology Support Unit, Harwell and the Energy Conservation Unit of East Kilbride

As can be seen, it is estimated by this quite authoritative committee that the scope for industrial energy savings is between 25 and 34 per cent of the total industrial fuel used.

This is, obviously, only a generalization. Some industries, particularly those where appreciable quantities of high grade waste heat are produced, lend themselves to waste energy recovery far better than others, which need mainly mechanical and electrical energy. Also, there is no doubt that some factories are already run along extremely energy-economical lines, while others are negligent in this respect.

AN ECONOMIC CONCEPT OF ENERGY

Although all forms of energy are expressed in the same units (joules, megajoules, gigajoules etc), the financial value of energy varies enormously with its grading. This means that electrical and mechanical energy are the most costly, followed by high grade heat energy.

At the other extreme, thermal energy which is only a few degrees above ambient has virtually no commercial value at all. This is, of course, the weakness of all these bright ideas about trying to equate the energy contained in solar radiation reaching the earth, or the heat content of geothermal fluids, with high grade energy obtainable from fossil fuel or nuclear power. For economic operation, one should never use energy at an appreciably higher grading rate than needed.

For example, electrical energy which has the highest energy grading of all, should be used for such purposes as mechanical energy generation, production of light, sound and very high temperatures in electrical furnaces. It is also the only suitable source of energy for electronic and similar purposes.

Electric fires, on the other hand, where electricity is basically used only for raising the ambient air temperature to about 20°C, are an extremely wasteful use for electricity. This applies both in the domestic and the industrial context. At present there is over-production of electricity at night and therefore some of the night electricity is sold off at cut price for space heating purposes. This is an inherent waste. It is far better to find more efficient methods of energy storage such as flywheels, compressed air or pumped water, in which there is far less thermodynamic irreversibility.

Under industrial conditions there are often several tasks which require energy, but which lie at different grading levels. It makes sense to use the waste heat from one process to serve the needs of another.

For example, glass works produce waste heat at between 400 and 500°C. This is quite sufficient to raise intermediate pressure steam for running back pressure turbines to produce electricity and heat in the form of low pressure steam at, say, 120°C. This, in its turn, could be used to evaporate moisture from agricultural products. The water vapour obtainable from such processes would probably be condensed to provide warm water at about 60°C. This could be employed for space heating or for the supply of heat to fish farms or greenhouses. If this is done, the original supply of high grade energy obtained by burning coal, oil or natural gas performs no less than four separate tasks:

The various glass constituents are melted and held above the solidification temperature (about 1500°C).

Medium pressure steam is used to produce electricity (500°C).

The exhaust steam from the back pressure turbine is used for crop drying (120°C).

The condensed water vapour is able to heat up water to be used for space heating, fish farms or greenhouses (60°C).

TABLE 1.3 THE UNITED KINGDOM ENERGY BALANCE (1976)

Total annual consumption of fossil fuel	Millions of coal equivalent 239.47	Petajoules (10^{15} J) 8553.5
Electricity industry (CEGB) used	68.26	2457.4
to produce the following electricity	21.56	776.3
rejecting as waste heat	46.70	1681.1
at an average temperature of 30°C to cooling towers, which represents an average fossil fuel to electricity conversion ratio of 31.585%		
Industry used	55.88	2013.3
plus purchased electricity of	11.02	396.9
yielding waste heat of	35 approx.	1270 approx.
at temperatures usually well over 30°C.		

Industrial evaporation

The glass industry represents the highest heat rejection grading with a temperature of up to 500°C. Pottery kilns provide waste heat at around 350°C. Most other industries reject heat at temperatures of 200°C and lower. Yet waste heat of this type is ideally suited for evaporating water from materials used in the paper and board industries, agricultural food industries, textile industries, etc. It has been calculated that the energy needed for industrial evaporation processes in which a total of 20.45 million tonnes of water is removed from wet products in Great Britain alone, amounts to the staggering quantity of 2.37 million tonnes of coal equivalent or 85.4 petajoules, worth at 1982 rates about £280 millions.

For all industrial evaporation processes, heat exchange media at a temperature of below 200°C are quite adequate. This means that waste heat rather than expensive prime fuel heat could be used with complete success.

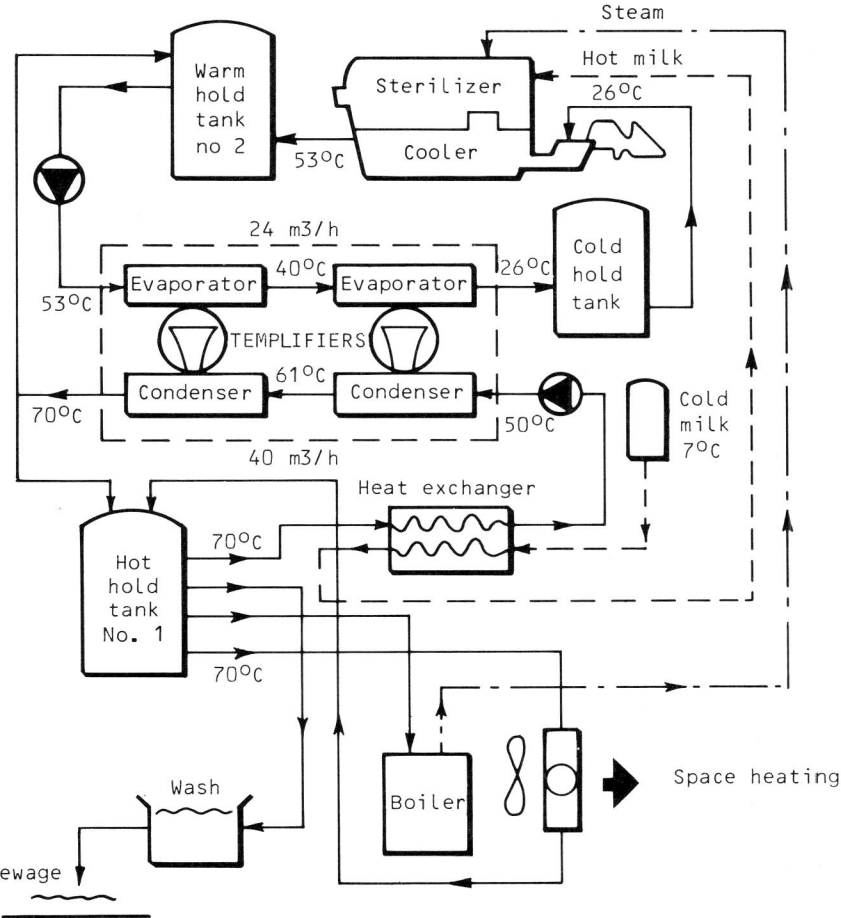

1.1 Simplified fluids diagram of the water saving and energy conservation scheme at the Unigate dairy in Walsall, UK

A typical case in point is the processing of milk into butter, cheese and milk powder. To achieve the 1976 annual production of 500 000 tonnes of these products in the UK 4.2 million tonnes of water have to be removed. It is in general necessary to employ three tonnes of low pressure steam for the removal of each tonne of water from milk. This is equal to 8 gigajoules (10^9 J) representing 222 kg of coal equivalent. As this quantity of coal or its equivalent costs about £25 (1982 prices), and one needs about one kg per kg of milk products produced, this gives an evaporation cost of about £130 per tonne of milk product. Yet low grade waste heat is all that is really needed, and if obtained from another source, it could cost very little.

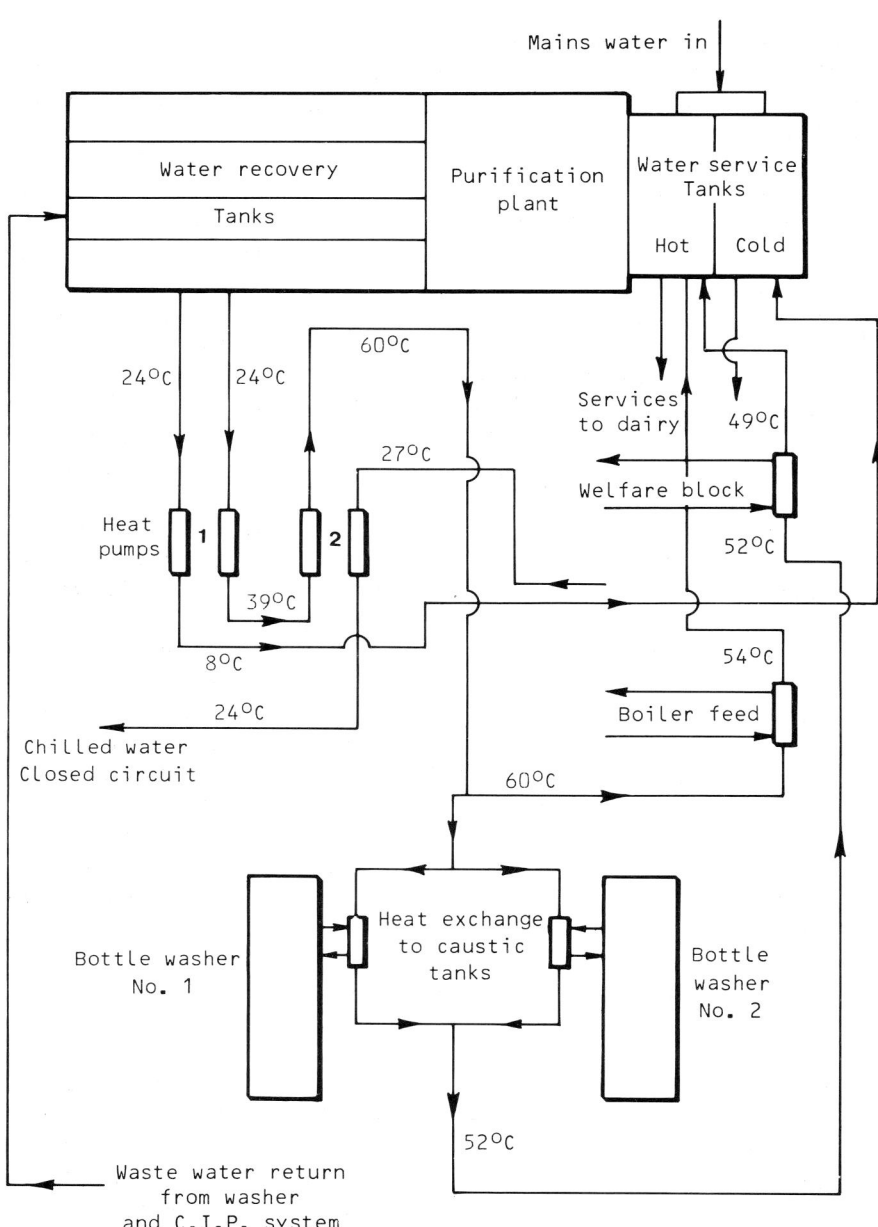

1.2 Schematic layout of water and heat recovery system, MMB Dairy, Bamber Bridge, Preston, UK

Agriculture and horticulture

These two industries employ huge quantities of low-grade heat for which waste heat produced by other industries would be extremely adequate.

Dried grass has a very much higher protein content than hay, which is grass dried by solar heat. The reason for this is that complex organic compounds are destroyed by the untraviolet rays which are constituents of sunlight. At the time of writing about 70 per cent of dried animal feed is imported from North America into Europe, largely because of lower fuel costs in the former. Yet the use of waste heat from industrial undertakings could lower costs sufficiently to make it a feasible operation in the UK and other northern European countries. Most crop dryers use air pre-heated to about 100–180°C. To dry vegetable matter to a satisfactory level some 120 GJ per tonne of dried material are required. This could be in the form of comparatively low grade waste heat from industrial processes.

There is at present considerable demand for heated greenhouse types of crops such as tomatoes, cucumbers and the like. The total value of such products this year (1982) is some £200 million, produced in greenhouse areas of over 2000 hectares of which about 1400 hectares are heated (1 ha = 100 m × 100 m). Because of competition from countries with natural sunshine it is difficult for producers in cold climates to compete effectively in growing salad crops if prime fuels have to be used. As the temperature grading of the warm water used in greenhouse heating is no higher than about 60°C, it would appear pointless to use prime fuels for generating this heat. In fact, all greenhouses require is secondary waste heat: ie heat which has already been degraded by serving as a medium for such purposes as crop drying. Fish farms too can use very low grade heat. Waste heat from several nuclear power stations (Hunterston is one example) is already used for this purpose.

Miscellaneous other users of low grade heat

In producing paper, pulp and board some 30 per cent of the energy used is for the drying process. 1½ tonnes of water have to be evaporated for each tonne of paper produced: drying is normally done by drums running at 200°C, providing exhaust air at between 70 and 120°C, which could be used to provide further waste heat. The current UK production rate for products of this type is about 4 million tonnes.

The food and drink sections of British industry use some 9 per cent of the total industrial energy consumed, or around 211 petajoules. In some sections, there is an extremely satisfactory balance between energy needs at different thermal levels. Most plants are already operating at good thermal efficiencies and are an example to other industries in this respect. The UK Thames sugar refinery, which produces 900 000 tonnes of refined sugar each year, is able to boast that the only thermal effluent from its factories is cooling water at 30°C. Some breweries are also extremely efficient, particularly those which operate total energy plants. In others there is considerable scope in the proper use of waste heat.

The petrochemical and general chemical industries

The petrochemical industry of Great Britain uses about 330 petajoules annually, and much of the heat rejected is at temperatures as high as 200°C. At present the petrochemical industry tends to make little use of the vast quantities of waste heat generated, which can amount to as much as 60 MW for a larger refinery.

Chemical industry varies enormously in terms of size, nature and factory location. In some plants adequate waste heat recovery is already practised. In others up to 100MW of useful waste heat at temperatures of 120°C and higher are still being wasted.

ECONOMICS OF USING WASTE HEAT

In all cases the utilisation of waste heat requires additional expenditure, both of a capital and a maintenance nature. This can be summarised as follows:

Waste gas ducts
Heat recovery equipment
Control and protection devices
Pipeline transmission.

The cost of waste heat recovery systems has to be compared with that of simple equipment which converts prime fuel into heat of the appropriate grade. In all cases the waste heat recovery plant will be far more expensive than that plant which uses prime fuel.

Let the cost of a waste heat utilization system be C_w £.

Let the cost of a prime fuel utilization system be C_p £.

Let the additional maintenance and labour cost for the waste heat recovery system be X £/year.

Let the fuel cost saved per annum be f £/year.

It is necessary in this case to allow for the likely average cost of fuel during the period of use of the plant as well as future maintenance and labour costs, as these will be subject to inflation. A simple pay-back period for a heat recovery system can be expressed as:

$$\frac{(C_w - C_p)}{f - X} \text{ years}$$

For most basic heat recuperator and regenerator systems pay-back periods of less than three years are normally required, owing to the short life of such equipment. For more complex power plant equipment

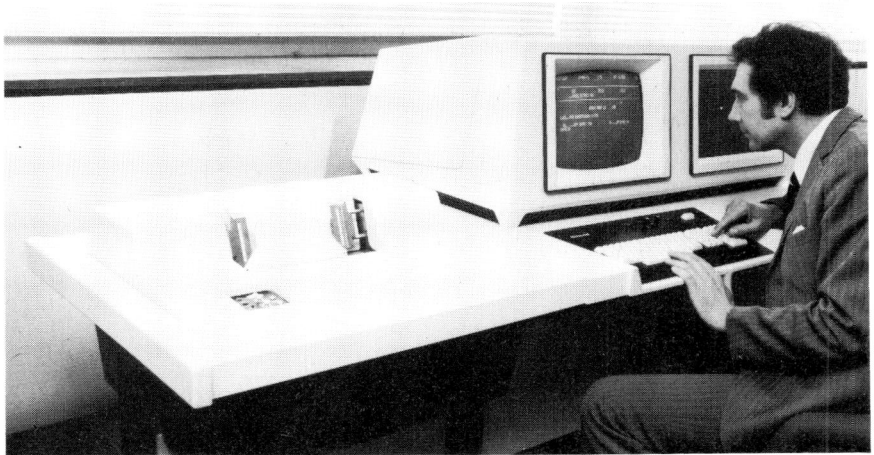

1.3 Operation of Honeywell computerized energy management system

pay-back periods of ten years and even longer are reasonably acceptable. It should be pointed out that these are very simple rule-of-thumb
criteria. On the other hand, as it is almost impossible to make genuinely
accurate prophecies as to future economic and political developments,
it is questionable whether there is really very much point in trying to be
particularly sophisticated in this respect.

Electronic control of heat supply

As was shown at the beginning of this chapter, the correct control of
heating appliances often has a more direct influence on the overall thermal economy of a building than many other bits of hardware. Furthermore, the installation of proper electronic control equipment is very
cost-effective.

Until very recently the only automatic control in most buildings was
an electro-mechanical time switch, able to turn the heating on and off at
predetermined times. This is not adequate with today's high fuel costs,
and far more sophisticated programmers are becoming economically
viable. In addition, modern microprocessor circuits can meet such
needs simply, easily and reliably.

The main requirements which have to be catered for by up-to-date
electronic control equipment can be summarised as follows:

Initial boost operation

The heating should be turned on as late as possible to prevent heating
up empty premises, yet the premises must be comfortably warm when
personnel arrive. This is normally done by an initial boost operation in
the morning. The magnitude of this boost and the time when it should
start are optimized electronically.

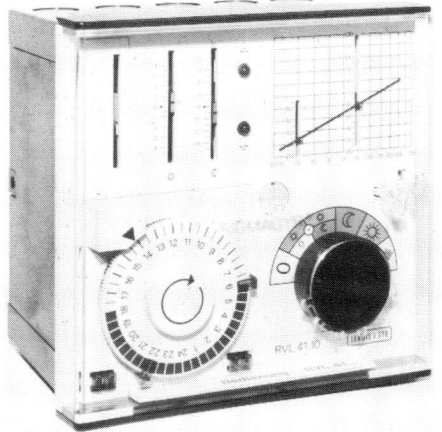

1.4 Modern electronic energy control device, Courtesy Landis & Gyr Ltd

1.5 Maximum demand electronic load controller. Courtesy Horstmann Ltd

Correct switch-on and switch-off patterns

The actual operation of the heating system is determined by the times of starting and finishing and the temperature levels required in each section of the building. The systems are pre-programmed for weekends, holidays, changeover to summertime and back.

External sensor devices

These are installed to warn of outside temperature drops, wind speeds, etc. They can be interpreted immediately by the computer so that it can modify the heating programmes to accord with such climatic changes. This gives a far better control of the interior climate than reliance upon internal thermostats, because of the very considerable delay caused by such factors as the thermal storage effect in the building structure.

Frost protection devices
These ensure that the temperature inside the premises can never fall
below +1°C either at night or during weekends and holidays by starting
up the heating system whenever there is any danger of frost damage.

Weather-compensated switch-on delay
A simple time-switch turns on the heating at a given time irrespective of
the weather outside. This is clearly not the most economical way of
operation: when the weather is mild the heating should be switched on
a good deal later than when it is very cold and there is a high wind. A
special switch-on delay mechanism works out the exact time and the
exact boosting necessary under different weather conditions.

Optimum cost analyzers
Gas and electricity costs sometimes vary according to the time of the
day when they are used. Optimum cost analyzers are used to evaluate
the optimum operation patterns for a heating system to make use of the
most favourable price structures.

Load shedding
In many systems the maximum demand must remain below a given
value. In such a case automatic load shedding is practised in the most
suitable way once there is a danger that the predicted maximum load
figure is likely to be breached.

Multi-boiler installations
In this case maximum operating efficiency is combined with minimum
fuel usage by the use of microprocessor boiler sequencers. This is
usually operated in conjunction with a frost control system. If any
boiler is switched off due to breakdown or for general maintenance, the
boiler which is out of commission is automatically moved in the shedding sequence to be the first off, and the other boilers take up sequential
jobs. This ensures that the defective boiler is only used under extreme
conditions.

After a boiler is switched off, the pump continues to run for 30
minutes to dissipate heat from the boilers into the building. This considerably assists in achieving optimum energy utilization, as it makes
use of heat which would otherwise be wasted.

Solar effect compensation
Many units have built-in solar effect compensation. Under direct sunshine, heating systems are switched off, and in some cases even
switched over to a cooling mode in which an air chilling circuit is substi-
tuted for the heating circuit. A system of this kind operates best with a
comprehensive heat pump installation.

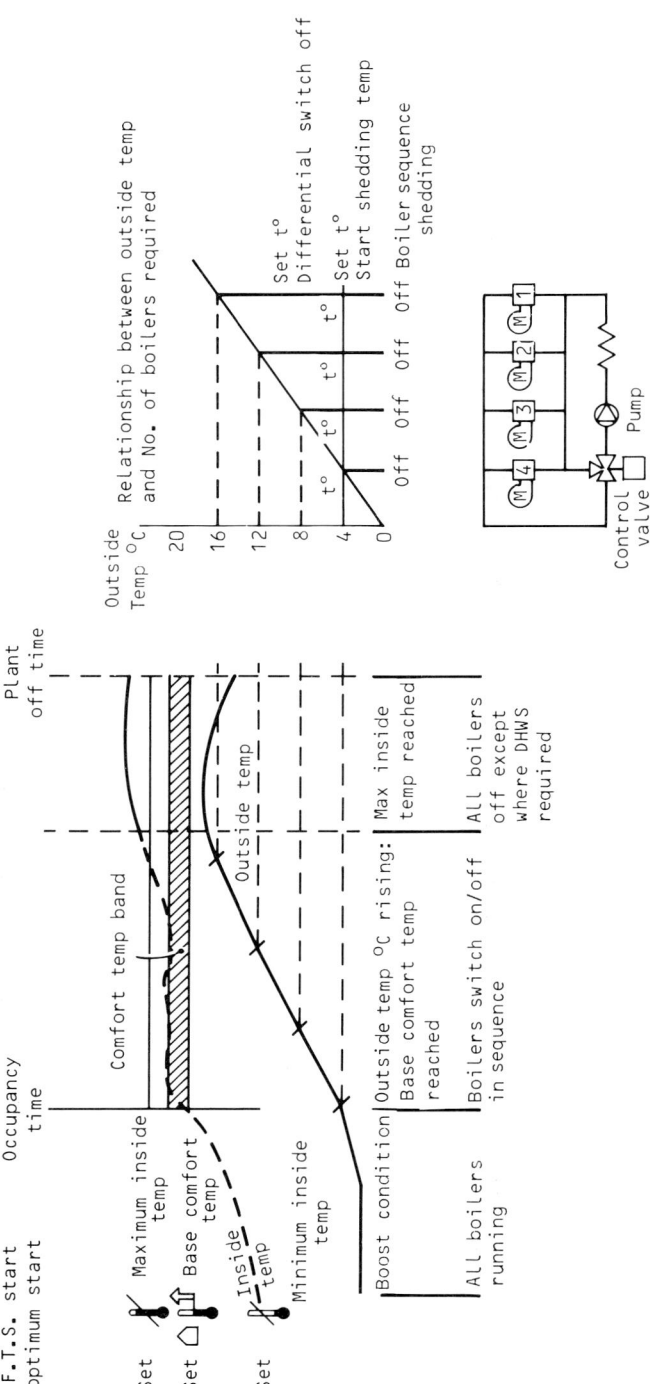

1.6 *Typical operation of boiler sequencer without electronic control*

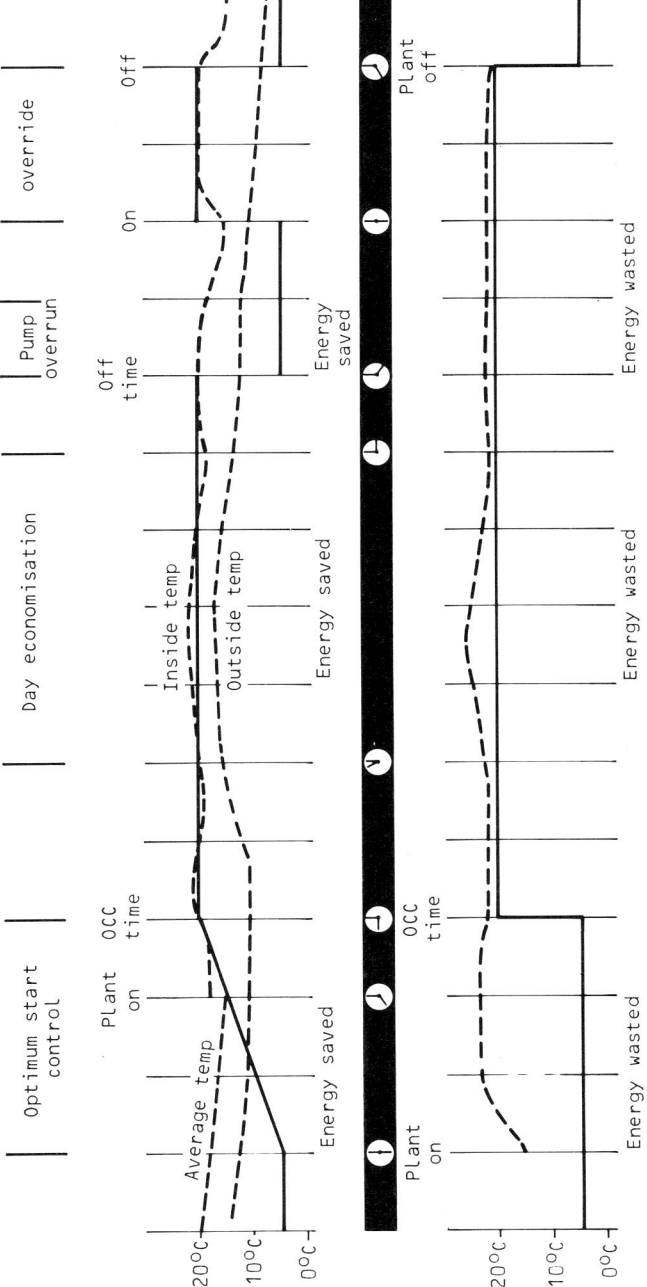

1.7 Operation of boiler sequencer with electronic control. Both courtesy JEL Ltd

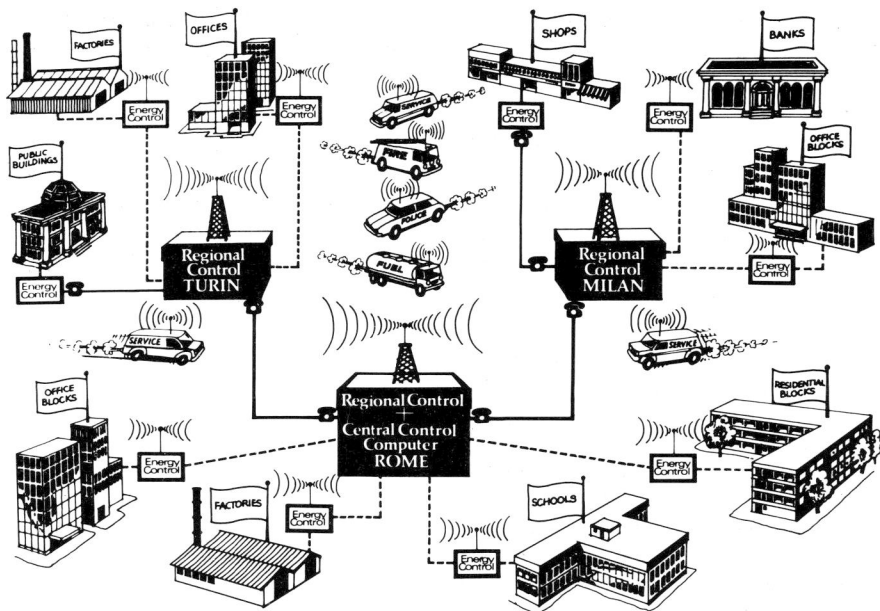

1.8 Centralized energy control in Italy. Courtesy JEL Ltd

The use of thinking microprocessor circuits

As already shown, the correct start of the heating operations is carried out by switching on the heating before staff arrive, and boosting the supply of heat somewhat to achieve the correct temperature at exactly the correct time. If the correct temperature is achieved either too early or too late, the microprocessor remembers this, and the following day makes an adjustment in the time of start and the intensity of the boost operation.

Equally, the computer switches off the boiler and pumps at the correct time so that the internal temperature does not fall below a set temperature at the time when the staff are due to go home. If the computer has either switched off too early or too late, this will be adjusted for the next day so that in future the switching-off takes place at exactly the right time. The electronic system is thus able to learn spontaneously by its own mistakes.

Safety systems

The control system is usually coupled to a safety system which can detect fire anywhere in the building instantaneously and capable of taking immediate action.

Miscellaneous functions

Numerous additional functions can be performed by the computer system, such as switching in other systems and units according to a time-switch programme. Common switching via a return network is often practised.

The various systems on the market differ from each other in the data shown on the console, the functions which can be performed and the system of control. Some of the newest systems enable the control of several factories to be carried out centrally using a telephone network.

This chapter summarises information supplied to me by various manufacturers of electronic control systems which differ from each other merely in details. There are, of course, numerous other manufacturers in this field, but I am indebted to:

Honeywell Ltd, Bracknell, Berks, UK.
Horstmann Engineering Products Ltd, Bath, UK.
JEL Energy Conservation Ltd, Stockport, Cheshire, UK.
Landys and Gyr Ltd, North Acton, London W3, UK.
Myson Products Ltd, Ongar, Essex, UK.
Rapaway Ltd, Shirley, Solihull, West Midlands, UK.
Redwood Industrial Products Ltd, Clevedon, Avon, UK.

References and further reading

1 A. McKillop, 'Global economic change and new energy', *Energy Policy* 9,4 December 1981, pp 229–235.
2 W. Laws, 'Waste heat as an energy source' *Energy World* 86, November 1981, pp 10–19.
3 J. Birks, 'Energy and new technology', *Energy World* 85, October 1981, pp 6–10.

SELECTED COMPANIES INVOLVED IN MANUFACTURE OF PRODUCTS

UK and Europe

AMF International Ltd, AMF House, Whitby Road, Bristol BS4 4AZ.
Appliance Components Ltd, Cordwallis Street, Maidenhead, Berks SL6 7BQ
Arcolectric Switches PLC, Central Avenue, East Molesey, Surrey KT8 0RF.
Asco (UK) Ltd, 2 Pit Hey Place, West Pimbo, Skelmersdale, Lancs.
Bestobell Mobrey Ltd, 190/196 Bath Road, Slough, Berks SL1 4DN.
Black Automatic Controls Ltd, Leafield, Corsham Wilts SN13 9SP,
Burgess and Co. Ltd, Easthampstead Road, Bracknell, Berks RG12 1NP.
Cistermiser Ltd, 176 Perimeter Road, Woodley, Reading, Berks RG5 4SN.
Danfoss A/S, Horsenden Lane South, Greenford, Middx UB6 7QE.
Digitron Instrumentation Ltd, Merchant Drive, Mead Lane Industrial Estate, Hertford, Herts SG13 7BH.
Drayton Controls Ltd, West Drayton, Middx UB7 7SP.
Eaton-Williams Group Ltd, Station Road, Edenbridge, Kent TN8 6EG.
Electronics Corporation of America (GB) Ltd, Sixth Floor, Tuition House, St. Georges Road, Wimbledon, London SW19 4XE.
Hamworthy Engineering Ltd, Fleets Corner, Poole, Dorset BH17 7LA.

JEL Energy Conversion Services Ltd, Edgeley Road Industrial Estate, Cheadle Heath, Stockport, Ches SK3 0XE.
Horstmann Gear Group Ltd, Newbridge Works, Bath, Avon BA1 3EF.
John Thurley Ltd, Ripon Road, Harrogate, North Yorks HG1 2BU.
Midland Temperature Control Systems Ltd, Chasetown Industrial Estate, Cannel Road, Chasetown, Walsall, West Midlands WS7 8JQ.
Myson Industrial Group Ltd, Industrial Estate, Ongar, Essex CM5 9RE.
National Utility Services Ltd, Carolyn House, Dingwall Road, Croydon CR9 3LX.
Oy Nokia AB, PO Box 44, 01511 Vantaa 51, Finland.
Parker Refrigeration Components Group, 69/71 Clarendon Road, PO Box 192, Watford, Herts WD1 1DQ.
Ranco Controls Ltd, Southway Drive, Southway, Plymouth PL6 6QT.
Randall Electronics Ltd, Ampthill Road, Bedford MK42 9ER.
H. Saacke Ltd, Fitzherbert Road, Farlington, Portsmouth PO6 1RX.
Sangamo Time Controls Ltd, Industrial Estate, Port Glasgow, Renfrews PA14 5XG.
Satchwell Control Systems Ltd, PO Box 57, Farnham Road, Slough, Berks SL1 4UH.
Sauter Automation Ltd, 165 Bath Road, Slough, Berks SL1 4AA
Smith Meters Ltd, 170 Rowan Road, Streatham Vale, London SW16 5JE.
Sopac-Jaeger Controls Ltd, 17 Invincible Road, Farnborough, Hants GU14 7QH.
Speedaire Supply Ltd, Cromford Road, Langley Mill, Notts NG16 4FL.
Sperryn Ltd, Delta Road, Parr, St. Helens, Merseyside WA9 2ED.
Summit Instruments Ltd, 471 Lichfield Road, Aston, Birmingham B6 7SP.
Tacotherm Ltd, 145 Oxford Street, London W1.
Teddington Controls Ltd, Daniels Lane, Holmbush, St. Austell, Cornwall PL25 3HS.
Teknigas Ltd, Charlwoods Place, East Grinstead, West Sussex RH19 2HY.
Transmitton Ltd, Smisby Road, Ashby-de-la-Zouch, Leics LE6 5UG.
Wilo (UK) Ltd, Unit 15, Olympic Industrial Estate, Wembley, Middx HA9 0XU.
Woods of Colchester Ltd, Tufnell Way, Colchester CO4 5AR.

United States

Action Instruments Inc., 8601 Aero Drive, San Diego, CA 92123.
Aegis Energy Systems Inc., 607 Airport Boulevard, Doylestown, PA 18901.
Alco Controls Division, PO Box 12700, St. Louis, MO 63141.
Altech Controls Corporation, 13955 Murphy Road, B1. 412, Stafford, TX 77477.
Atlantic Energy Technologies Inc., 10 Keith Way, Hingham, MA 02043.
Cam-Stat Inc., 11833 W. Olympic, Los Angeles, CA 90064.
Controlled Energy Systems, 1240 NE 175th Street, Seattle, WA 98155.
CSL industries, 11040 Santa Monica Boulevard, Los Angeles, CA 90025.
Encon Systems Inc., 504-C Vandell Way, Campbell, CA 95008.
Energy Microsystems Inc., 9026 Hague Road, Indianapolis, IN 46256.
Fuel Computer Corporation of America, 419 Whalley Avenue, New Haven, CT 06511.

Garetano Associates Inc., 1304 Motor Parkway, Hauppauge, NY 11788.
International Energy Management Div. of Schindler Haughton, 671 Spencer Street, PO Box 780, Toledo, OH 43695.
LESP Inc., PO Box 1699, Atlanta, GA 30371.
Microcomputer Ventures Inc., 3857 N. High Street, Columbus, OH 43212.
Microcontrol Systems Inc., 6579 N. Sidney P1 Milwaukee, WI 53209.
National Energy Corporation, 1820 Shelburne Road, Burlington, VT 05401.
Pacific Technology Inc., 235 Airport Way, PO Box 149, Renton, WA 98055.
Power Management Systems Inc., 12 S. 12th Street, PSFS Bld., Philadelphia, PA 19107.
Ranco Controls, 8115 U.S. Route 42 N., Plain City, OH 43064.
Reliance Time Controls Inc., 1820 Layard Avenue, Racine, WI 53404.
Scientific-Atlanta Inc., One Technology Parkway, Atlanta, GA 30348.
SSAC Inc., PO Box 395, Liverpool, NY 13088.
Staefa Control Systems Inc., 2481 San Leandro Boulevard, San Landro, CA 94577.
Temperature Systems Inc., 159 Amony Street, Manchester, NH 03102.
Trimax Controls Inc., 1180 Miraloma Way, Sunnyvale, CA 94086.
Veeder Root Co., 70 Sargeant Street, Hartford CT 06102.
Vertrex Corp., 2807 E. Madison Street, Seattle, WA 98112.
Xencon Co., 150 Mitchell Boulevard, San Rafael, CA 94903.

2 Radiant heating equipment

Heating industrial buildings such as factories, workshops, storage halls and aircraft hangars is often a problem due to their considerable internal heights. In many cases such buildings need floor to ceiling heights of well over 10 m. Yet human beings, who are the only contents which need to be kept warm, do not operate above a height of 2 m. The same problems are also found, incidentally, with quite a few non-industrial buildings, such as churches, indoor swimming pools and sports halls.

Normal convection heating uses a fuel to warm up air from the external ambient temperature to room temperature, and then blows it in. Heat losses are partly due to convection, ie when the hot room air is changed and decanted to the outside, and by conduction through walls, windows and roof. As hot air rises, we find that a situation develops in which the dead space well above the heads of the people who should be comfortable is actually the zone with the highest temperature. This means that heat losses through the roof are unacceptably high. Furthermore the temperature of the air is a good deal higher than the temperature of the floor and walls, which is neither economic nor conducive to the maximum comfort conditions.

As will be shown later on, comfort is optimised when floor temperatures are higher than air temperatures. Such conditions are never found with convection heating but are a general feature of radiant heating practice.

Radiant heating systems
Radiant heating systems can generally be subdivided into several specific types:

Panel heaters
Panel heaters fed by either steam or hot water can be used either in the ceilings, walls or floors. Walls are not a very popular location. As far as floor heating is concerned, one uses embedded heating tubes in the concrete slab and allows the surface to radiate heat upwards.

Naturally such a system has its limitations because surface temperatures of floors must be kept below about 40°C. It is, however, a useful way of using low grade thermal energy.

Panel and coil heaters are often mounted in ceilings and radiate heat downwards. Again the use of either hot water or steam limits the oper-

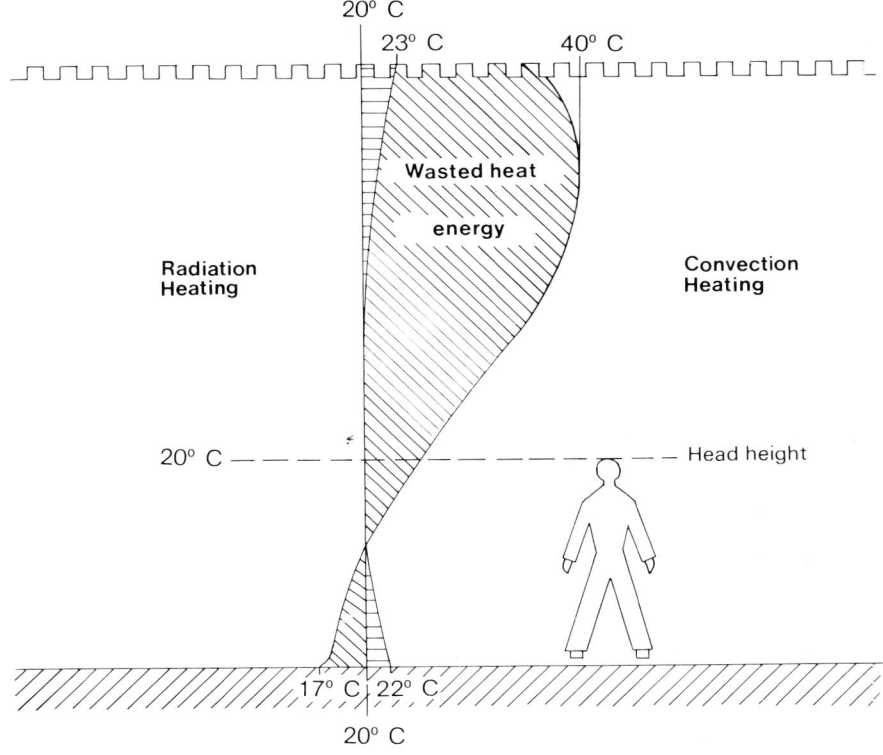

20° C

23° C   40° C

Wasted heat

energy

Radiation
Heating

Convection
Heating

20° C — — — —   — — — Head height

17° C 22° C

20° C

2.1 Vertical heat distribution patterns using radiant or convected heating.
Courtesy Kubler GmbH

ating temperature. Surfaces for such radiant panel heaters usually consist of vitreous enamelled steel which has a top temperature limit of about 260°C.

Other radiant steam and hot water heaters simply use pipe coils embedded in the plaster of the walls. Plaster has an emissivity of about 0.9 compared with 1.0 for a perfect black body, and is thus a satisfactory radiation surface.

Hot water and steam heated coils have a very much lower radiation efficiency than either the radiant panel heaters or the tubular heaters because of the comparatively low temperature of the radiant heat surface. From the user's point of view, however, such systems have two enormous advantages:

1 Such systems are the only ones which can be permitted when flammable liquids and gases are present. Typical uses are for workshops where paint spraying takes place, or in fuel stores. Care has to be taken to check that the flash-point of the flammable vapour is at all times above the temperature of the radiant heating surface.

2.2 Hot water overhead radiant panels. Courtesy HCP Ltd

2 Radiant heaters of this type can use low grade waste heat which would otherwise have to be disposed of expensively. Of particular value in this respect is underfloor heating, where the coils can be fed with water at only around 40°C. From the energy economy point of view this is most valuable as water at such a low exergy (energy grading) level would otherwise be a virtually valueless commodity.

Radiant heaters using electricity

These can either be of the oil filled variety with a black external tubular body, or they can employ a silica structure which is raised to incandescence by an embedded resistance wire.

The former is used as a low temperature black body radiator and can be employed where flammable products are present. As with steam or hot water radiant heaters, care must be taken that the flashpoint of any flammable vapours in contact with the tubular heater is always well above the temperature of the heater. The usual precautions must be taken to avoid sparking at contact breaker points. As the surface temperature of such radiant heaters is comparatively low, radiation efficiencies are not high. The incandescent silica tubes operate at much higher temperatures, namely around 700–850°C.

As in all radiant heaters, the Stefan Boltzmann equation applies, which is given as:

Heat given off = $5.673 \times 10^{-8} A e ((T_1)^4 - (T_2)^4)$ watts

where A is the area of the radiation surface in m^2
where e is the emissivity (dimensionless)
(black body = 1, ceramics around 0.9)
T_1 is the temperature of the radiating surface in kelvins (°C + 273.15)
T_2 is the temperature of the receiving surface, again in kelvins (°C + 273.15).

Obviously the amount of radiation per unit surface area of electrically heated silica heaters is much higher than that of black body heaters which are seldom operated at temperatures exceeding about 140°C.

According to EURA (Energy Users Research Association) the following cost levels (October 1982) apply for electricity as against various gaseous fuels in terms of pence per useful GJ of heat output.

Fuel	Heat output cost (Oct 1982) pence/GJ
Electricity (on peak)	10.06
Electricity (off peak)	4.31
Natural gas	3.19
Propane	4.59
Butane	4.04

These figures apply to purchases in excess of 10 000 GJ/annum. It can therefore be seen that the use of on-peak electricity for heating cannot be defended economically as such a fuel is about two and a half times as expensive as alternative gas fuels. On the other hand, if off-peak electricity can be used to operate radiant heaters, there is very little cost differential between electricity and gas.

Electricity, then, has a real advantage as auxiliary equipment, for electricity supply is far cheaper than the auxiliary equipment (flues, gas storage, control valves) needed for the combustion of gas.

One of the reasons why night electricity is so reasonably priced is that much of it comes from nuclear sources. The cost of electricity from nuclear reactors is far lower than from plant using conventional fossil fuels.

As far as comparisons can be made between gas fired and electric radiant heaters, and radiant heaters using steam or hot water as a heat source, EURA gives the following costs (October 1982) for various boiler fuels:

Boiler fuel	Heat output cost (Oct 1982) pence/GJ
Fuel oil (35 seconds)	4.51–5.99
Fuel oil (200 seconds)	3.90–4.89
Fuel oil (950 seconds)	3.23–4.28
Fuel oil (3500 seconds)	2.91–4.01
Coal	1.97
Industrial coke	3.07
High quality blast furnace coke	3.13
Large South Wales foundry coke	3.53

So it can be seen that when either steam or hot water are used as a heating medium, there seem to be few real cost advantages (except in the case of coal) to counterbalance the greatly increased cost of having to instal a boiler system and pipeline network, if one can use either off-peak electric power or gas as a heating medium instead.

Plaque type gas fired radiant heaters
In these gas, which may be either natural gas or manufactured propane or butane, is burned against a ceramic structure called a plaque, which is raised to incandescence and radiates heat.

TABLE 2.1

Temperature of radiating surface in °C	Heat radiation from emittor in W/m²
100	612
150	1 259
200	2 181
250	3 447
300	5 132
350	7 321
400	10 106
450	13 585
500	17 866
550	23 063
600	29 299
650	36 703
700	45 413
750	55 574
800	67 339
850	80 869
900	96 332
950	113 903
1000	133 767

The radiation emitted from such heaters is of a very high order because ceramic materials have an emissivity of about 0.9 and, as already shown, radiant heat emission varies with the fourth power of the surface temperature. **Table 2.1** gives the actual amount of heat radiated by radiating surface with emissivity 0.9, assuming that the surroundings are kept at a temperature of 20°C.

It can therefore be seen that plaque type radiant heaters operating at close to 900°C, emit about 9½ times as much radiant heat per unit of radiation surface as tubular heaters operating at about 400°C, and between 40 and 80 times as much heat as some electrically or steam heated appliances operating between 150 and 200°C.

Obviously from a theoretical point of view the actual quantity of heat radiated per unit of useful fuel burned must be the same, whatever the surface temperature of the radiator used. Manufacturers of plaque type overhead gas radiators, however, claim that from an overall radiation point of view overhead plaque radiators show an advantage of 2.57 per cent over equivalent black body radiators operating at temperatures of around 400°C as against 850°C for the plaque radiators: it is claimed that plaque radiators have a lower convection heat loss than tubular heaters (K. E. Parkes of Parkinson Cowan GWB, in *Building Services & Environmental Engineer* April 1980).

The following test data were obtained from the British Gas Corporation when testing a Parkinson Cowan Flamrad Model 606 G plaque type heater:

Gas flow	$0.559 \text{m}^3/\text{s}$
Gas calorific value	38 MJ/m^3
Input heat	21 216 kW
Plaque temperature	850°C
Plaque area	0.124 m^2
Plaque emissivity factor	0.89
Useful radiative output	10 000 kW
Radiative efficiency	$\dfrac{10\,000}{21\,216} = 47.13\%$

In an attempt to arbitrate in the heated battle between the manufacturers of plaque type overhead radiant heaters and black body tubular installations, the writer's opinion is that plaque type heaters may indeed give a fractional improvement in radiation heat output per useful GJ in the fuel as against metal tubular heaters: but it is doubtful whether the figure can be as high as 2.57 per cent. This is based upon an emissivity factor for tubular heaters (quoted by a plaque heater maker) of 0.8 and a reflector emissivity factor of 0.35. If one assumes a more reasonable factor of 0.85 for the tubular heater and 0.2 for the reflector, differences in efficiency between the two heaters would be of a much lower order.

Plaque heaters can be made much smaller than tubular heaters: an

FORM 1: PLAN AND SIDE SECTION

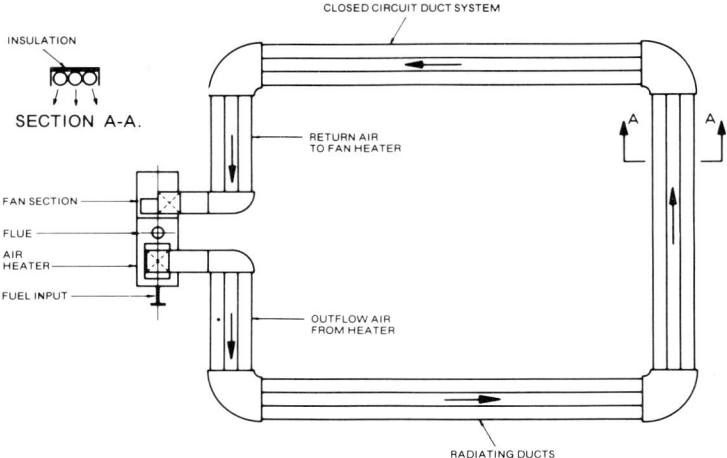

FORM 2: PLAN AND SIDE SECTION

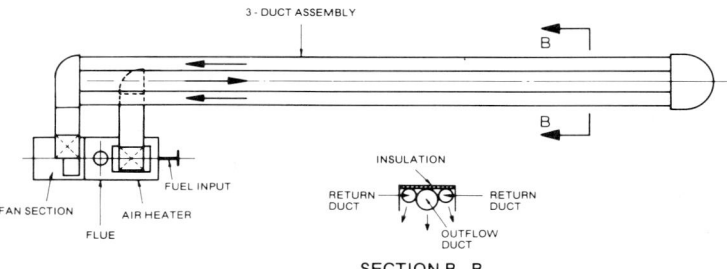

FORM 3: PERSPECTIVE DRAWING

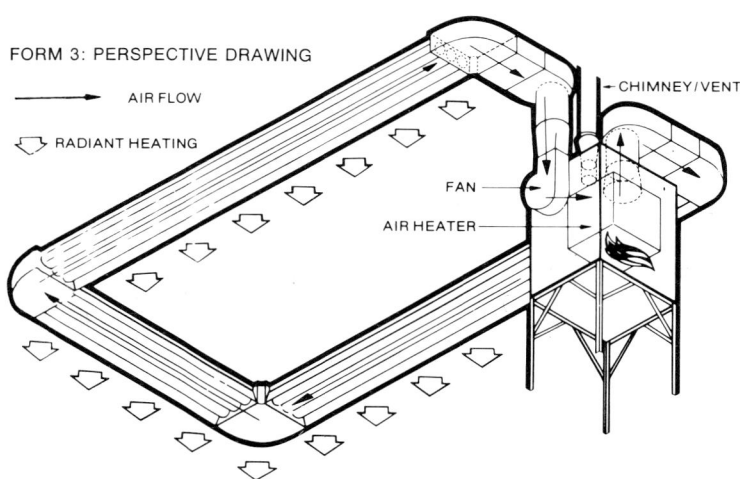

2.3 Three views of radiant tube heating. Courtesy Radiant Tube System Ltd

advantage where space is at a premium. On the other hand, ceramic plaques are much more sensitive to rough handling than steel tubes, which are the basic features of tubular radiant heaters. Under certain conditions plaques could also constitute a fire risk.

Plaque type radiant heaters can be run with either natural gas or with propane and butane, and can be positioned either horizontally or at an angle of 45° to the horizontal.

Tubular radiant heaters

These are customarily in the form of a U-tube with external diameter of about 100 mm, with either a black or rough grey surface. The tube is heated internally by gas burners, and the heat is spread uniformly along its surface by an electrically driven fan. The system is cheap, safe and easy to apply. The surface temperature of the tubes usually ranges between 150°C and 500°C.

The higher the temperature of the heater tube, the greater is the fraction of heat which is transmitted by radiation. This is due to the difference in formulae dealing with convection and radiation heat transfer.
Let Q_c = heat transferred by convection in watts and let A be surface area in m^2.
For convection heat transfer one uses the formula:
$h = 2.5 (\Delta T)^{1/2} W/m^2 K$ for a top surface of a heated body and
$h = 1.32 (\Delta T)^{1/2} W/m^2 K$ for the bottom surface of a heated body, ie if one adds the two a reasonable h value is equal to:
$3.82 (\Delta T)^{1/2} W/m^2 K$ where ΔT is the temperature difference between the heated body and the outside air in kelvins
$Q_c = h A (\Delta T)$ watts.
For radiation heating one uses the Stefan Boltzmann equation to evaluate the heat which is being transmitted.

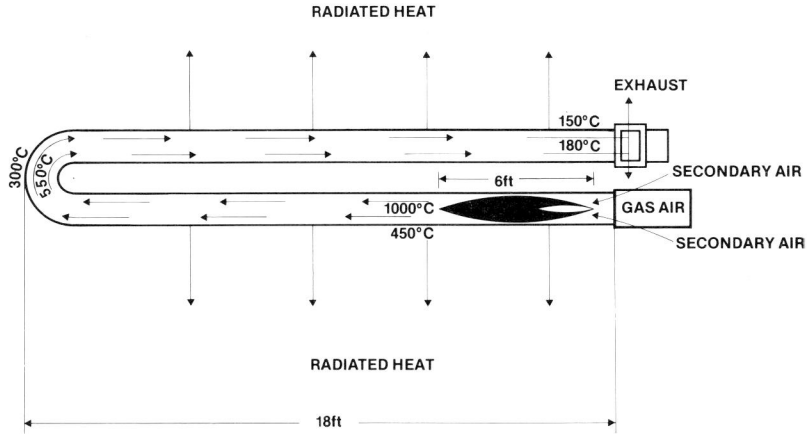

2.4 Design of typical radiant heater. Courtesy Grayhill-Westcott Ltd

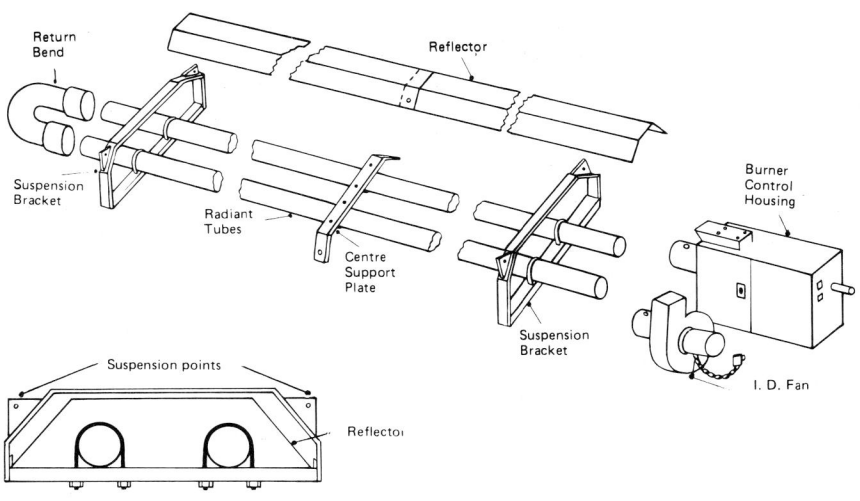

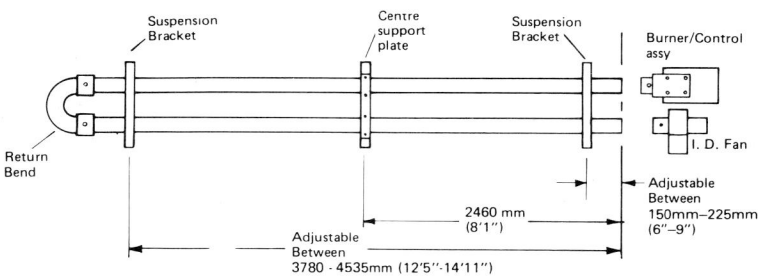

Layout of Gas Controls (Viewed from top of control housing)

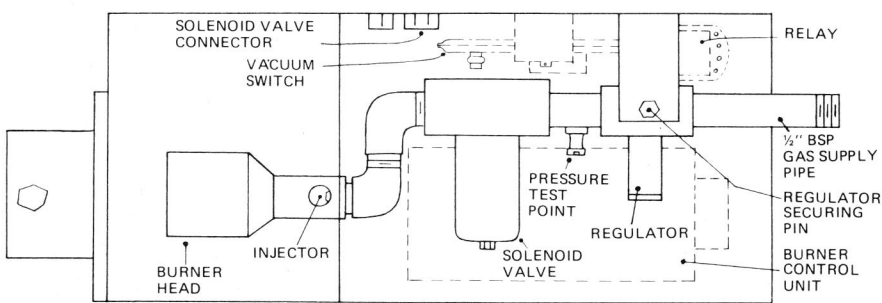

2.5 The Ambirad tubular radiant heating system

Most radiant heaters employ shiny metal reflectors to direct the radiant heat where it is needed. If we assume the absorptivity = emissivity of the radiant heater tube to be equal to 1, while the absorptivity of the floor or other receiver surfaces is equal to e, we can write that the heat transferred by radiation from the radiator is equal to:

$Q_r = 5.673 \times 10^{-8} \, f \, A \, e \, (T_1^4 - T_2^4)$ watts

where f is a factor of 0.9 included to make allowances for absorption of heat by air, geometric factors, etc. T_1 and T_2 are the absolute temperatures of the heater body and the flooring surfaces respectively

Let us assume a floor temperature and air temperature of 18°C and examine how the ratio of radiated and convected air is affected by the surface temperature of the radiant heater.

Let e = 0.85.

TABLE 2.2 CONVECTION AND RADIATION HEAT EMISSION

Surface temperature of heater °C	K	Q_c kW/m²	Q_r kW/m²	Q_r/Q_c (dl)
150	423	1.709	1.078	0.63
175	448	2.122	1.437	0.68
200	473	2.554	1.861	0.73
225	498	3.000	2.358	0.79
250	523	3.459	2.936	0.85
275	548	3,930	3.603	0.92
300	573	4.414	4.367	0.99
325	598	4.909	5.239	1.07
350	623	5.414	6.227	1.15
375	648	5.928	7.341	1.24
400	673	6.451	8.592	1.33
425	698	6.983	9.990	1.43
450	723	7.523	11.547	1.53
475	748	8.072	13.274	1.64
500	773	8.627	15.184	1.76

As can be seen from **table 2.2**, the higher the surface temperature, the greater the proportion of radiated heat to convected heat. In practice, due to air shielding on top of the radiant tubes, the ratio of radiated heat to convected heat is somewhat more favourable than the figure above, which derives from purely theoretical considerations. Experiments carried out by Dr J. K. Maund of the Chemical Engineering Dept. of the University of Aston gave the results for four types of commercial gas fired radiant heaters, all intended to be mounted at a high level.

When the units are inclined at an angle of 45° to the horizontal, convection heat from each unit increases appreciably. This does not apply to units which operate at high temperatures, such as plaque heaters and some tubular ones.

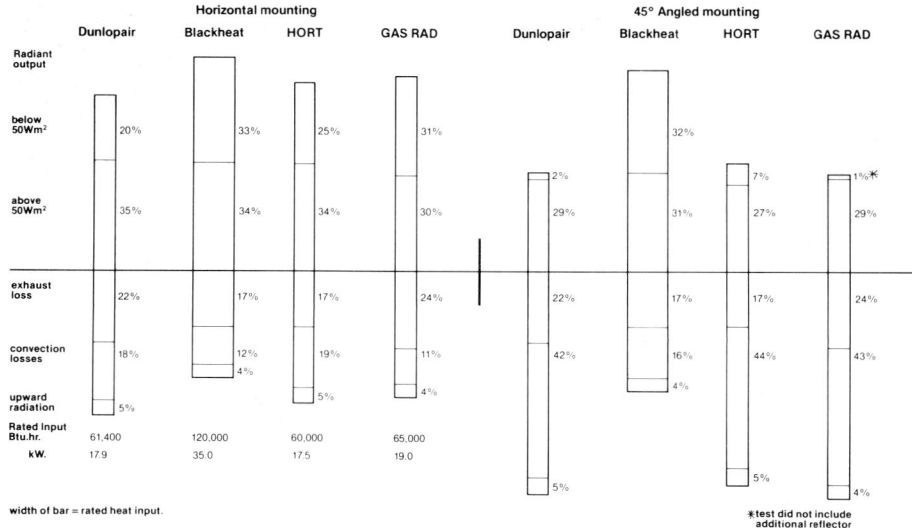

2.6 *Comparative radiant output of heaters tested. Courtesy Dr J. K. Maund, University of Aston, Birmingham.*

High grade radiation is defined as a radiation of intensity exceeding 50 W/m^2 at a distance of 2.5 m from the centre line of the tube emitters, while any figure less than this is calculated as low grade radiation.

When operating radiant heaters one seeks to keep the convection heat output as low as possible, since convection raises the temperature of the air close to the radiant heater tube. Because tubular radiant heaters are placed high up and there is no facility for circulating the hot air downward, convected heat is therefore wasted heat. For most commercial radiant heaters the practical limit is about 70 per cent of the calorific value of the gas used in the form of actual radiant heat.

When low output units are placed at an angle, convection heat output increases at the expense of radiant heat, although some types counteract this by means of specially designed reflector units.

In addition to the gas input, all tubular radiant heaters require single phase 240 v AC electric power through a standard 13 amp socket. Fan motors have an average power consumption of around 45 watts.

In most cases it was found that the maximum U-tube temperature, which corresponds to the maximum heat output of the burner, was at a distance of 1 m from the burner end.

It is best to use designs where the temperature is kept at over 400°C along the first tube of the system by using internal baffles to improve convection of heat from the hot gas to the tube walls. As has already been shown, the higher the temperature at which one can keep the surface of the tubular heat radiators, the greater the percentage of heat transmitted in the form of radiation. However, it has been found that

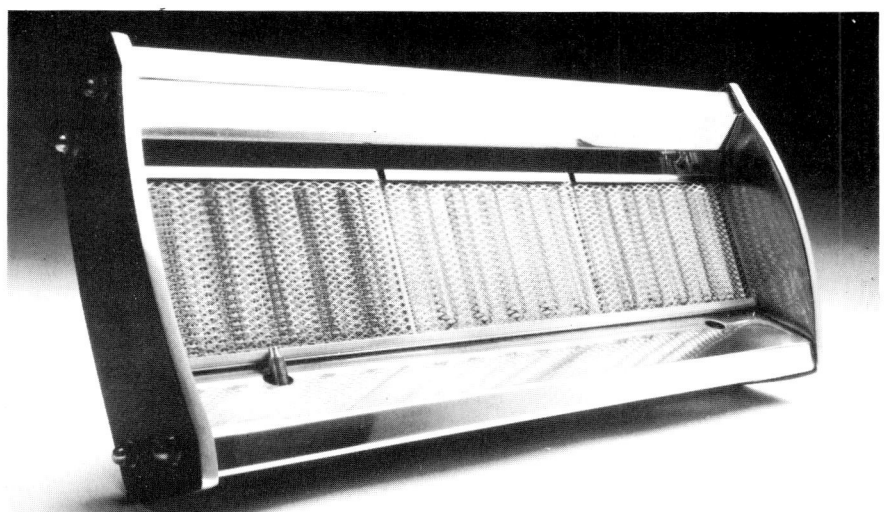

2.7 The Radiant overhead gas fired plaque heater. Courtesy Radiant Superjet Ltd

there is a limit of 500°C for the steel tubes, as temperatures above this can cause the materials to become suspect. It has been found that rough unpainted steel surfaces soon become coated with a thin blue-black film which has an absorptivity/emissivity close to unity for infra-red radiation.

Reflectors
An essential part of a tube system is the reflector: in nearly all cases this is a shiny aluminium shield fixed above and around the radiant heating tubes, which reflects virtually all the radiant heat downwards. Although heat is radiated from the heating tubes towards the reflectors, their absorptivity coefficient is so low that only a small fraction is absorbed by them. They are mainly heated by convection.

In practice it is found that the reflector surface temperature is limited to 150°C when the heating tube surface is at 450°C. This considerably reduces convection heat transfer from the reflector surface upwards. It is possible to use alternative materials such as stainless steel, galvanised steel or chrome plated steel strip, but these have been found to be less efficient in use than aluminium because their surface finish deteriorates and reduces the initial reflectivity. Once the radiant heat has been received by the floor, machines, walls and other surfaces, these act as secondary heat emitters and redistribute the radiant heat to other sections of the building. This ensures that hidden areas are also heated, since radiant heat, like visible light, can only travel in a straight line.

Comfort conditions

As comfort is a subjective phenomenon it is difficult to assess by proper scientific means. However, some attempts have been made to do precisely that.

When hot air convection heating is used, circulating air is passed through the heater to be warmed up. It is then mixed with the room air. As there is a boundary layer of stagnant air against the wall and floor surfaces, heat transfer to these cannot be completed and in consequence such surfaces have a temperature some 3–4°C lower than the air itself. Heat exchange by radiation takes place between the human body and these cool surfaces, and this induces the feeling of a lower temperature than actually exists.

When radiant heating is used, the air only receives a fairly small portion of the heat; that part convected from the radiant heat source and also from the wall and floor, in the form of secondary convection. The floor, walls and other objects in the building are consistently at a higher temperature than the room air and therefore the subject receives heat radiation from these secondary radiators as well.

As far back as 1936 Dr Bedford developed a complex formula relating the equivalent temperature from the point of view of comfort conditions. Using a Fahrenheit scale for temperature and expressing wind velocity in terms of feet/second he obtained the following equation, which to the mind of the writer seems a bit complex for the expression of subjective experience.

$$t_e = 0.522\ t_a + 0.478\ t_w - 0.01474\ (v(100 - t_a))^{1/2}$$

which reduces under internal conditions to:

$$t_e = 0.522\ t_a + 0.478\ t_w$$

where t_a = air temperature in °F

t_e = effective temperature in °F

t_w = wall temperature in °F

It is probably not too inaccurate to substitute Centigrade values and rewrite the Bedford formula as:

$$t_e = 0.52\ t_a + 0.48\ t_w$$

Let us take the two hypothetical conditions:

1 The air temperature is 22°C and the wall temperature is 18°C.

$$t_e = 0.52 \times 22 + 0.48 \times 18 = 20.08°C$$

2 The air temperature is 18°C and the wall temperature 22°C

$$t_e = 0.52 \times 18 + 0.48 \times 22 = 19.92°C$$

It can therefore be seen that the comfort conditions under these two circumstances are virtually identical.

Several experiments have been carried out by the ECRC at Capenhurst on the influence of asymmetric radiation on comfort conditions. People were invited to sit in a chair in a corner of a test room for 45 minutes under radiant heat emission from overhead radiators and then invited to express the degree of pleasantness or unpleasantness on a scale of 1 to 7.

The concept of the vrt or vector radiant temperature was developed, which gives the average surface temperature of one half of a room minus that of the other. The units employed are kelvins. No increase of mean discomfort vote was found at a vrt of 20 K and in consequence this figure was taken as an acceptable limit. Other workers in the field such as McNall and Biddison developed a calculated vrt of 24 K without increasing discomfort conditions, yet some put a limit to as low as 10.5 K.

Obviously comfort conditions can not as yet be evaluated to any degree of accuracy. It seems, however, that there is little evidence that people are more comfortable under conditions of high air temperature and low floor and wall temperatures with the absence of radiant heat emission, than when air temperatures are kept low but heat is being supplied by radiant emitters and the surfaces surrounding the subject are at a temperature higher than the ambient air.

Reduction of heat losses through roof

As has already been shown, convection heating of high factory halls and similar buildings, which requires the heating up of the entire volume of air inside them, is extremely wasteful in fuel. Internal heights can be quite considerable yet there is really no need to keep the air warm at a height of more than 2 m. Yet with convection heating the opposite seems to take place. As hot air rises, the air layers closest to the ceiling are at their warmest.

Practical measurements carried out in Germany by the Kubler Organisation have shown that with hot air heating the pattern of temperature distribution shown in figure 2.1 occurs. While the air temperature at head height, (around the 2 m level) is 20°C, the temperature of the floor is only 17°C. Yet at a distance of about 0.5 m below ceiling level temperatures can go up to 40°C, being only slightly below this temperature at actual ceiling level. Heat losses through a wall or ceiling interface can be expressed by the equation:

$$Q = UA (\triangle T)$$

where   Q is the heat loss in watts

U is the heat transfer coefficient in W/m^2 K

$\triangle T$ is the temperature difference in kelvins, and

A is the area in m^2

It can therefore be seen that heat losses due to conduction through interfaces can be reduced considerably if internal temperatures can be kept down. This is especially true of factory roofs, which need to incorporate appreciable glass areas and hence are badly insulated.

When one achieves an equal degree of comfort by using ceiling mounted radiant heaters, air temperatures close to the ceiling are very much reduced. The German report gives a figure of 23°C as against 37°C with air convector heating.

Let us assume that the factory roof consists partly of simple asphalt on concrete with a U-value of 3.4 W/m² K and partly of single glazing with U-value 8.3 under the exposed conditions prevailing with skylights.

Reducing the air temperature close to the ceiling from 37°C to 23°C therefore saves:

$14 \times 3.4 = 47.6$ W/m² in the case of the roof and
$14 \times 8.3 = 116.2$ W/m² in the case of the skylight.

If one assumes 1280 hours of operation per annum, this works out at a saving of 219.3 MJ/m² for the roof and 535.4 MJ/m² for the window area. There are also considerable savings in heat energy at the upper wall levels. These savings become higher if the quality of insulation is low.

Convection heat losses

Let us assume that the specific heat of air at constant pressure equals 29.22 J/mole K. Then at an average temperature of 18°C the specific heat of 1m³ of air equals:

$$\frac{29.22 \times 1000 \times 273}{22.414 \times 291} = 1223 \text{ J/m}^3 \text{ K}$$

Let us assume that by using radiation heating, one can reduce the average air temperature by 6°C, a figure commonly quoted in the literature. Let us consider a factory building 120 m × 50 m × 12 m in height, requiring 2 air changes per hour. Heat savings can be calculated to amount to:

$2 \times 120 \times 50 \times 12 \times 6 \times 1223$ J/h = 1057 MJ

which works out at 1352.5 GJ of heat over the 1280 hour heating season. Even when cheap North Sea gas is used, which costs about £1.80 per useful GJ, this still represents a saving of £2434.50 over the heating season. To this, of course, we have to add the heat savings due to conduction against the ceiling, skylights and upper walls. With radiant heating floor temperatures are higher and therefore one must expect some increase in conduction losses. However, even with badly insulated floors, U-values there are a good deal lower than with roofs, skylights, sidings and windows.

Most manufacturers claim that heat savings obtained when substituting radiant heating from overhead tubular radiators for conventional convector heaters are of the order of 50 per cent without sacrifice of comfort conditions. As was shown above one can achieve comfort conditions with lower air temperatures and higher floor temperatures (as produced by overhead radiant heaters) equal to those obtainable when using much higher air temperatures but low floor temperatures (which go with warm air convection heating).

The author is indebted to Grayhill Wescott Ltd of Poole, Dorset, for the following cost balance for running a radiant heater unit with input of 35 kW (120 000 Btu/h) in order to heat an area of 186 m² (2000 ft²) of factory floor with a radiation intensity of an average of 131.7 W/m². Such a figure would be appropriate to an average engineering workshop or garage. Premises in which light sedentary work is being carried on would have to employ roughly 1.5 times this radiation intensity.

Gas costs about 17.7p per 100 MJ in the UK.

Heater uses 120 MJ per hour	21.24 p
Allowance for 45 watts of electricity to drive fan	0.76 p
gving an actual running cost per hour of	22.00 p
Assuming a 40 hour week and a 32 week heating year with heater having to be used 70 per cent of the total 1280 hours involved	896 hours
Total cost per annum of running the heater is	£197.12

This gives a total annual cost equal to 107.7 p per m² (9.86 p per ft²) of floor area

Installation

Radiant heaters should be installed at a minimum height of some 4 m above the floor, preferably higher.

The radiation intensity in W/m² is inversely proportional to the perpendicular square of the distance between the heater and the floor area to be heated. The matter is somewhat complicated by the facts that the radiant heaters do not have a completely uniform surface temperature and also that part of the heat is first radiated upwards and then reflected downwards by the shiny metal reflectors. Further complications are the cut-off angle of the reflectors and the penumbra effect: at certain angles only the heat given off by one section of the pipe is actually reflected. All this means that the actual amount of heat received by floor, machinery and humans is somewhat variable. However, since secondary radiation is emitted by these primary receivers of radiant heat, this does not detract from overall comfort conditions.

Heat input related to activity

The actual radiant heat required varies considerably with the type of work which is being carried out. In workshops where heavy labour is being performed or where only background heating is required, as in heavy machine workshops, machinery stores, motor engineering workshops and the like, the heat input should be calculated on the basis of 150 W/m².

This should be increased to about 200 W/m² in workshops where less actual manual effort is required, or where operatives are more on the move. The highest heat input is required in premises where work is sedentary. Examples are fine assembly areas for electronic or other instruments, machining areas in the clothing industry, drawing offices and the like, particularly where the bulk of the labour employed is female. Under such circumstances a radiation intensity of around 250–300 W/m² is required to give adequate comfort conditions. To calculate the area of factory space which can be covered by one heater we simply divide the effective radiant input of the heater by the heat requirements in W/m² ie:

$$\text{area covered} = \frac{C}{HR}$$

where C is the capacity of the heater in watts and HR is the heat requirement in W/m²

Example: Assuming the heater input equals 35 kW or 35 000 watts and the degree of heating needed is 200 W/m², we can work out that such a heater is sufficient to cope with an area of:

$$\frac{35\,000}{200}\ m^2 = 175\ m^2\ (1\,880\ ft^2)$$

Heater height

This depends largely upon the design of the shielding reflector. Few radiant heaters operate properly at a height below 3.5 m above the ground, as the heat output is too uneven. If it is absolutely necessary to mount the radiant heaters lower than this, it becomes essential to employ diffusers or to use angular mounting.

Let us assume that at a height of x metres between the radiant heater and the floor the radiation intensity equals RI W/m² and that the heater is able to cope with an area of A m². Then an increase in height from x to y reduces the radiation intensity by $\left(\frac{x}{y}\right)^2$ and increases the area covered to $A\left(\frac{y}{x}\right)^2$.

SELECTION CHART: RADIANT HEATERS

Type of equipment	Advantages	Disadvantages	Appropriate applications
Panel heaters fed with steam or hot water	Fireproof Use low grade energy No flues needed	Poor radiant heat emitters	Areas where flammable vapours are present
Electricity heated radiators	Fireproof No flues needed	High fuel cost	When very cheap electricity is available (nuclear, hydro-electric)

SELECTION CHART: RADIANT HEATERS—*continued*

Type of equipment	Advantages	Disadvantages	Appropriate applications
Incandescent radiant heaters	Very high radiation efficiency	Ceramic components easily damaged; care has to be taken that no flammable materials come into contact	Industrial applications where maximum radiation and minimum convection heating is required
Tubular radiant heaters	Good radiant heat emission Long-lasting with low maintenance cost	Have to be placed high up to be effective Reasonably fireproof except in presence of very flammable materials	High factory halls, churches, sports halls, indoor swimming pools

References and further reading

1 T. Bedford, *The warmth factor in comfort at work*, Report of Industrial Health Research Board, 76, London, 1936.

2 F. A. Chrenko, 'Threshold intensities of thermal radiation evoking sensations of warmth', *J. Physiology* 173, 1964, p 1.

3 T. W. Oppel and J. D. Hardy, 'Studies in temperature sensation', *J. Clinical Investigations*, 16, 1937, p 517.

4 F. A. Chrenko, *Radiant heat and thermal comfort electricity and space heating*, Blackie 1965, pp 216–223.

5 I. D. Griffiths and D. A. McIntyre, 'Radiant heating and comfort', *Building Systems Engineer* 40, June 1972, pp 32–43.

6 D. A. McIntyre, 'Overhead radiation and comfort', *Building Systems Engineer* 44, January 1977, pp 226–234.

7 R. M. E. Diamant, *Insulation Deskbook,* Heating and Ventilation Publications Ltd, 1977.

8 J. K. Maund, *A comparison of four gasfired radiant heaters*, University of Aston Report, April 1979.

9 K. E. Parkes, 'Radiant heater efficiency', *Building Services & Environmental Engineer*, April 1980, p 28.

SELECTED COMPANIES INVOLVED IN MANUFACTURE OF PRODUCTS

UK and Europe

Armca Specialities Ltd, Armca House, 102 Beehive Lane, Ilford, Essex IG4 5EQ.

Bahco Ltd, Beaumont Road, Banbury, Oxon OX16 17B.

F. H. Biddle Ltd, Newtown Road, Nuneaton, Warwickshire CV11 4HP.

Colt International Ltd, New Lane, Havant, Hants.

Covrad Ltd, Sir Henry Parkes Road, Canley, Coventry CV5 6BN.

Dunham-Bush Ltd, Fitzherbert Road, Farlington, Portsmouth, Hants.
Dunlop Energy Engineering Ltd, Somers Road Industrial Estate, Rugby, Warwickshire.
Frenger Ceilings Ltd, Gerrards House, Station Road, Gerrards Cross, Bucks.
Gas-Fired Products Ltd, Claydon, Ipswich, Suffolk.
Gibbons Bros Ltd, PO Box 20, Lenches Road, Brierley Hill, Staffs.
Grayhill Wescott Ltd, 52 Nuffield Road, Poole, Dorset.
Hamworthy Engineering Ltd, Fleets Corner, Poole, Dorset BH17 7LA.
HCP Ltd, Unit 1, Block D, Castleham Industrial Estate, St Leonards-on-Sea, East Sussex TN38 9NU.
Joule Manufacturing Ltd, Unit 1, Padgets Lane, Moons Moat, South Industrial Estate, Redditch,Worcs.
Kiloheat Ltd, 21/22 Vestry Estate, Sevenoaks, Kent.
Mather and Platt Ltd, Park Works, Manchester M10 6BA.
Minikay Ltd, Manstead House, High Road, Chadwell Heath, Romford, Essex.
Moducel Ltd, 165 King Street, Fenton, Stoke-on-Trent ST4 3ES.
Myson Group Ltd, Old Wolverton, Milton Keynes MK1 5PT.
Novenco Ltd, Tundry Way, Chainbridge Road, Blaydon-on-Tyne NE21 5SN.
Parkinson Cowan Ltd, PO Box 4, Burton Works, Dudley, Worcs.
Phoenix Burners Ltd, 34–44 Tunstall Road, London SW9 8DA.
Radiant Superjet Ltd, Clapgate Lane, Woodgate, Birmingham B32 3BP.
Radiant Tube Systems Ltd, Swift House, Morgans Lane, Tooley Street, London SE1.
Schwank Ltd, 62 Sunningdale Road, Sutton, Surrey SM1 2JS.
Spiral Tube and Components Ltd, Osmaston Park Road, Derby.
Standard and Pochin Ltd, Evington Valley Road, Leicester LE5 5LS.
Trane Europa, Marketing Division, Europa 104, 94532 Eungis Cedex France.
Wanson Ltd, 7 Elstree Way, Boreham Wood, Herts WD6 1SA.

United States
Airtex Corporation, 2900 N. Western Avenue, Chicago, IL 60618.
Aitken Products Inc., 566 N. Eagle Street, Geneva, OH 44041.
American Infra-Red Inc., 11840 Edlie Road, Detroit, MI 48214.
Bio-Energy Systems Inc., 221 Canal Street, Ellenville, NY 12428.
Black Body CTX Products, 1526 Fenpark Drive, Fenton, MO 63026.
Cox Mfg Inc., 108 W. Second Street, Ridgeville, IN 47380.
Detroit Radiant Products Co., 1297 Terminal Avenue, Detroit, MI 48214.
Energy Conservation Inc., 2685 E. 79th Street, Cleveland OH 44104.
Lambert Industries Inc., PO Box 207, Lakeville, MN 55044.
Perfection Infra-red Products Co., PO Box 40, Waynesboro, GA 30830.
Roberts-Gordon Appliance Corp., PO Box 44, Buffalo, NY 14240.
Russ Sawdo Sales Inc., 1800 Broadway, Buffalo, NY 14212.
Shadyside Industries Raycon Div., 19th & Union Street, PO Box 218, Bellaire, OH 43906.
Shelley Radiant Ceiling Co. Inc., 456 W. Frontage Road, Northfield, IL 60093.
Solaronics Inc., 704 Woodward Avenue, Rochester, MI 48063.

3 Total energy from prime movers

Electricity is the most convenient general form of energy. It can be converted at a very high efficiency into mechanical energy, sound energy, light energy and heat energy. Equally, it can be transported enormous distances without undue loss and directed to exactly the position where conversion into other forms of energy is needed. Its production from basic fuels is, however, most inefficient.

In the chain of reactions:

Fuel → heat → mechanical energy → electrical energy

the weakest link in all cases is the conversion from heat to mechanical energy. In even the most perfect systems the efficiency of conversion is limited by Carnot's equation:

$$\text{Efficiency} < \frac{T_2 - T_1}{T_2}$$

where T_2 and T_1 are the highest and the lowest temperatures of the system in degrees absolute, respectively.

In practice T_2 is represented by the maximum gas temperature achieved inside the diesel engine, gas engine or gas turbine, or the maximum saturation steam temperature inside a steam turbine. T_1 is represented by the temperature of the exhaust gases of a diesel or gas plant, and by the temperature of the condensate steam in the case of a steam plant.

As exhaust gases in gas turbines and similar equipment emerge at around 770 K while the operating temperatures are seldom much over 1050 K, the thermodynamic limitation of efficiency of the heat/energy conversion process with such typical gas turbines is:

$$\frac{1050 - 770}{1050} = 26.67 \text{ per cent}$$

It is, of course, possible to design gas turbines with lower exhaust temperatures and much higher combustion chamber temperatures. Some of the recently developed closed cycle gas turbines have combustion temperatures as high as 900°C (1173 K), corresponding to a Carnot efficiency of 34.4 per cent. After discounting some of the other thermodynamic losses inherent in the system the maximum conversion efficiency of the chemical energy contained in the fuel into mechanical energy tends to be well below 30 per cent.

Steam turbines operate under rather different conditions. Modern sets use steam at temperatures of up to 580°C and exhaust at temperatures as low as 30°C. Steam, however, deviates considerably from the ideal gas laws. Even though theoretically a steam turbine should give an efficiency of

$$\frac{(580 + 273) - (30 + 273)}{(580 + 273)} = \frac{853 - 303}{853} = 64.5 \text{ per cent}$$

in actual fact even the most modern steam plants seldom if ever exceed about 36–37 per cent in efficiency.

Nor is there much likelihood of any great improvement in the future. The lower limit is already set by the high vacuum operated on the exhaust side, which can hardly be improved upon. Better construction materials may make it possible to raise the boiler pressure fractionally, but this is unlikely to improve the overall efficiency of power generation by more than a few per cent.

Total energy
In order to achieve better fuel economy than that obtainable when using prime movers on a fuel to electric energy cycle alone, it is often advisable to use the TOTAL ENERGY path. In this one separates those operations which require high-grade electrical energy, such as mechanical work, lighting, electronics, high temperature furnaces and others, from those which only need low-grade heat energy. The latter include space heating, industrial process heat for drying and similar operations, refrigeration employing absorption plant, and the like.

If the exhaust temperature of the prime mover is high enough, as in gas turbines, gas engines and diesel engines, the power generating efficiency of the prime mover is hardly affected at all by the process of heat extraction.

Steam turbines are different because of the very low exhaust temperatures normally employed. If one wishes to extract heat at a useful temperature level some sacrifice of electricity generating capacity is normally inevitable.

General economic discussion
Central electricity generation using low grades of oil or coal, water power and particularly nuclear power, is a fairly economic proposition, in spite of an average thermal efficiency of only 27 per cent. This is because the fuels used are very cheap. Central electricity generation, however, suffers from other cost disadvantages:

a The large bureaucracy which is found particularly with state owned and administered undertakings.

b The provisions of step-up transformers to lift the voltage from the approximately 20 kV of production to 400 kV at which power is normally transmitted.

c The provision of transmission lines, which are particularly costly when for environmental and other considerations they have to be placed underground.

In certain cases where, for example, supply lines have to be very long in relationship to the actual power consumed, or where frequent maintenance or replacement is to be expected (high winds, marine atmospheres, etc.) transmission charges become very high indeed.

d The provision of step-down transformers, needed for lowering the voltage from 400 kV to either 240 v or 120 v, at which the bulk of electricity is consumed.

e A considerable amount of electric power is converted into low grade heat in the step-up and the step-down transformers, and also during actual transmission through the power lines.

Comparative transmission costs of electricity and other forms of energy are given in **table 3.1**.

Except when the central power generation authority uses CHP turbines for district heating purposes, the waste heat produced during the actual electric power generation process is normally wasted entirely, being extracted either by cooling towers or by sea- or fresh-water coolers.

The alternative to purchasing electricity from a central utility, combined with the buying of prime fuel for space heating and process heating purposes is so-called TOTAL ENERGY operation. The consumer makes his own electricity using small scale plant and utilizes the waste heat produced. On the face of it, this seems a very attractive proposition, but in actual fact there are a number of pitfalls. It is essential that any total energy scheme is properly costed and assessed in terms of its suitability for specific cases.

TABLE 3.1 TRANSMISSION COST OF ENERGY

(assuming equal quantities of energy, transmitted for distance of about 100 km)

	Index (dimensionless)
Electricity at 400 kV (including transformer charges)	100
Oil (including pumping stations)	27
Natural gas	31
Water at temperature 150–200°C using single pipeline (including pumping station, heat loss)	48
Hot water using flow and return line, (including heat losses, pumping costs)	67

The following are the advantages and disadvantages of total energy operation as against purchase of electricity from a central authority, plus prime fuel for space and process heating.

Advantages
a Independence from grid system, when for some reason or other (hurricanes, plant breakdown, strikes) supply is erratic.
b Much higher utilization factor for fuel than with central electricity generation. This is partly counteracted by the lower fuel cost which utilities pay.
c Savings in administration and transmission costs which are inevitably passed on to the consumer in higher power charges.
d Savings in plant cost. Mass-produced small scale total energy equipment is cheaper per unit output than one-off large scale plant. The additional plant investment cost is naturally passed on to the consumer in the form of higher power charges.

Disadvantages
a When total energy is used, the developer has to find large capital investment for his plant. When he is connected to the power utility this is the problem of the utility and not his.
b Extra investment is needed as the capacity required expands.
c Need for skilled personnel in total energy installations. There is no need for them when the consumer is simply connected to a centralized supply.
d Production costs per unit rise considerably as the load shrinks, unlike connection to central generation plants, where this is not the case.
e The operator of a total energy installation has to concern himself about such matters as noise, smoke emission, fuel storage and pollution, as against the user of centrally produced electricity, who has no such worries.
f Costs of central power generation are reasonably predictable, while with total energy schemes a lot has still to be taken on trust. Costs cannot be forseen with any accuracy.
g The flexibility of consumption inherent with purchase of electric power from a central utility is lost. Plant overheads have to be borne even when power requirements are low. With purchase of power from central utilities one only pays for the power which one uses.
h Fuel costs for operating total energy plant are high. They can only be justified if one can get a far higher overall utilization figure for the energy contained in the fuel than is represented by the electric power extracted. As, in general, one gets between 2 and 3 units of waste heat for each unit of electric energy produced, it is necessary to have a steady demand for such large quantities of comparatively low grade heat energy.

MAIN PRIME MOVERS EMPLOYED

STEAM TURBINES

Steam turbines are the main prime movers used in very large plant, such as that used for the production of electricity by public utilities and very large factories.
Definitions of various types of steam plant in use:

Condensing turbines

In these the inlet steam is at about 530°C and 110 bars pressure: well above the critical temperature for water, which is 374.15°C. This high-pressure steam passes through the turbine blades of the high pressure set, and is usually reheated to get rid of any water condensed. It then passes through the low-pressure set, to be exhausted at a vacuum. This is of the order of 4 kPa (0.04 bars), at which pressure the steam temperature falls to about 30°C. To maintain a vacuum, this very low pressure and temperature steam has to be condensed. This is done by either a normal fresh water or seawater cooler, or using a cooling tower arrangement. A well-arranged set of this type can produce up to 36 per cent of the net calorific value of the fuel supplied to the plant, in the form of electricity. The condensed water is, however, at too low a temperature to be of any value.

Back pressure turbines

In these the exhaust steam comes off at a much higher temperature and

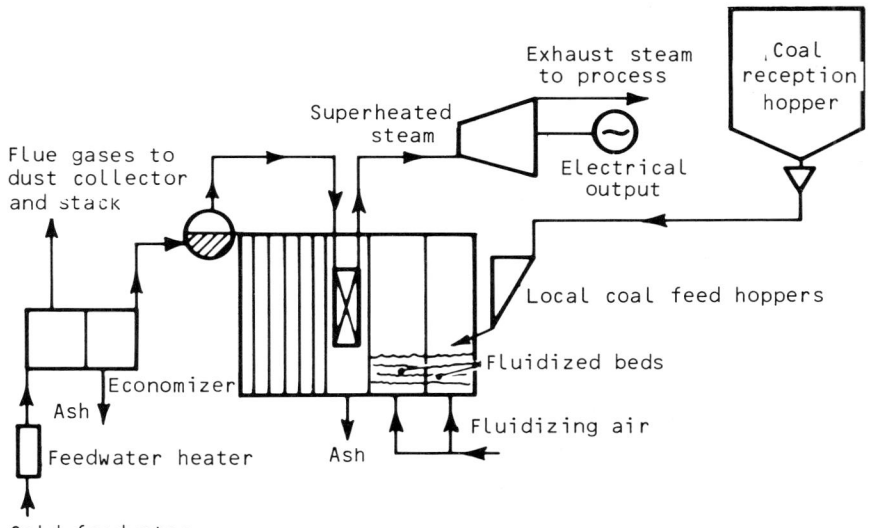

3.1 Total energy system using an A. P. E. Allen steam turbine as prime mover

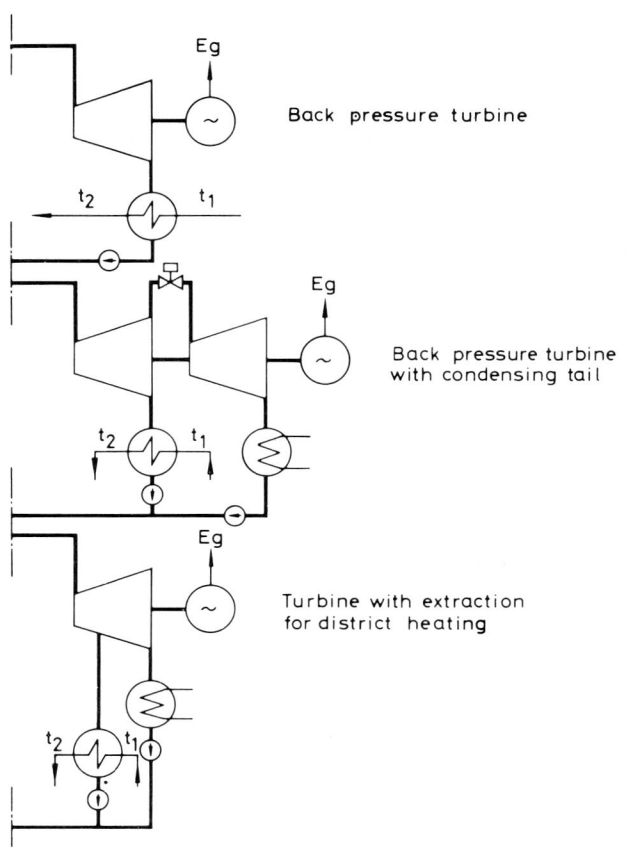

Back pressure turbine

Back pressure turbine
with condensing tail

Turbine with extraction
for district heating

Steam turbines for combined production of
electricity and district heating

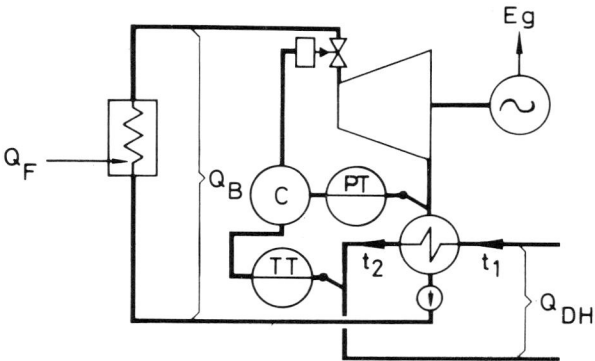

Back pressure turbine

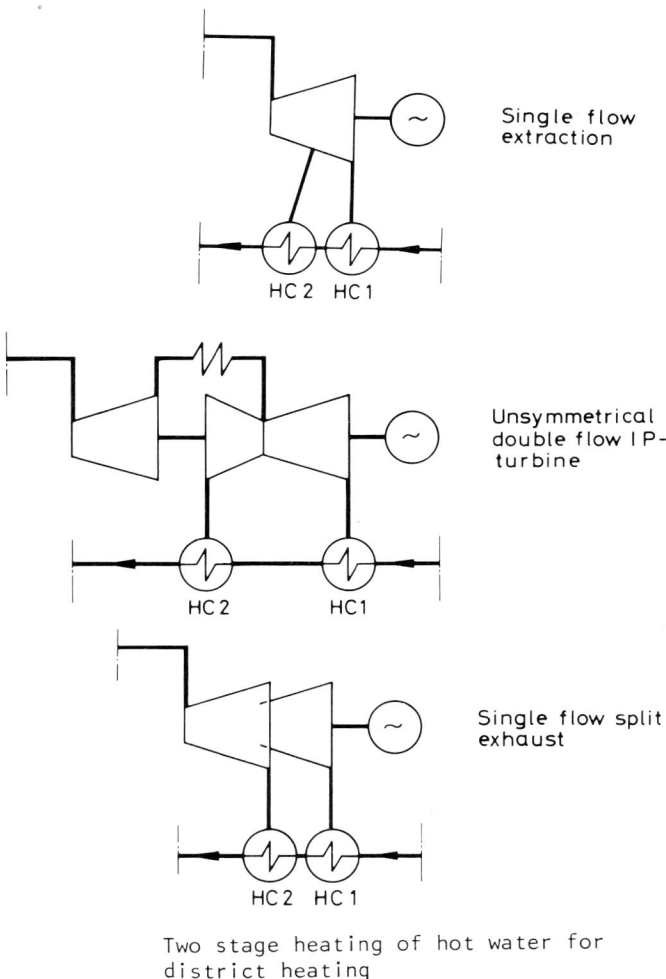

Single flow
extraction

HC 2   HC 1

Unsymmetrical
double flow I P-
turbine

HC 2   HC1

Single flow split
exhaust

HC 2   HC 1

Two stage heating of hot water for
district heating

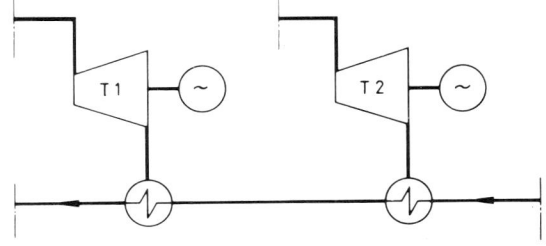

T 1   ~   T 2   ~

Two back pressure turbines in series on
the hot water side

*3.2 Various types of steam turbine used in total energy systems. Courtesy
Stal-Laval.*

pressure than in condensing turbines. The temperature of the exhaust steam is chosen in relationship to the use which is intended for this heat. It is normally within the range of 80—120°C.

Although the electricity generating efficiency of back pressure turbines is well below that of equivalent condensing turbines, the combined heat and power generating efficiency is a maximum and can go as high as 85—90 per cent of the net calorific value of the fuel used.

In general, some 1.5 to 2 times as much heat is produced as electricity and furthermore, this ratio cannot be varied except within a very narrow range.

Back pressure turbines are used by industries which use very large quantities of steam, such as firms involved in the drying of agricultural products or sugar manufacturers. Their electric power demand is rather modest in contrast to their steam demand.

Back pressure turbines are also used in some very large European district heating networks, to maintain base load heating supply. Often such turbines are fed with steam raised by the combustion of refuse or other unconventional fuel.

Pass-out turbines

In order to obtain the highest possible output of electricity when power demand is high but heating demand is either very low or not-existent, one uses a combination of condensing and back pressure turbine. This is called the pass-out turbine or the ITOC turbine (intermediate take-off condensing). There are numerous different designs of these. Basically such turbines can operate on a 100 per cent condensing cycle if so desired, while permitting the extraction of steam at various positions and temperatures. Such a system is very energy-economic. One only sacrifices one unit of electricity for each 6–9 units of heat supplied, depending upon the temperature level at which the steam is extracted and condensed.

Pass-out turbines are widely used to operate large-scale European city heating schemes. The turbine plants are very large, often several hundred MW in size. When operated at peak demand periods they have an efficiency of electricity production virtually the same as that offered by a conventional condensing turbine plant.

At periods when both power and heat demand are low, the turbine is operated with maximum steam take-off. The hot water produced by the extraction of the latent heat from the steam is stored in the pipeline network and in specially constructed large, highly insulated tanks called accumulators.

Industrial applications of steam turbine total energy systems
a Back pressure operation with surplus steam valves
It is almost impossible to balance the steam/power ratio with an uncon-

trolled back pressure turbine. If one seeks to use such a turbine for total energy operation it is essential to employ valves able to blow off steam to a dump condenser. This has the advantage that the steam is not lost, even though its latent heat is. It is condensed to water which can be reused without having to pass through a treatment process. Schemes of this type should only be employed if steam demand is steady most of the time, although slight fluctuations cannot be guaranteed against.

b Combination of back pressure turbine and purchased electricity
Provided one can obtain the agreement of the public power utility company, it is often advantageous to run a back pressure plant in conjunction with purchased electricity. Under such circumstances the turbine is operated so that the steam input is entirely governed by the process steam demand. Electricity is considered simply as a by-product and the total power demand is covered by purchasing sufficient extra power from the public utility to match the requirements. One obtains the best of both worlds in this way.

c The use of an accumulator in conjunction with a back pressure turbine
This system involves the incorporation of a steam storage vessel able to balance steam demand and output. It is a system which can only be used if less electricity is required overall than is produced by the prime mover in supplying the entire steam demand. The back pressure turbine is operated in such a way as to match the power demand. Steam is fed from the back pressure side of the turbine and from auxiliary boilers into a set of accumulators. The exact quantity of process steam needed is then extracted from this source.

d The use of a pass-out turbine
This system is used whenever process steam demand is small in relationship to electricity demand, or varies appreciably over the day. The system is particularly suitable for cases where waste fuel or other waste heat is to be utilized, for which there is no immediate process heat demand. It is necessary, however, that the additional power generated can be used or sold.

e Combination of condensing turbine and back pressure turbine
This is an alternative to the pass-out turbine, which is used when there is a clear difference between summer usage and winter use. In summer, when heating demand is low, one utilizes the condensing turbine alone, while in winter, when space heating is also required, the back pressure turbine is used. The back pressure turbine is operated in such a way that its output is governed by the steam requirement rather than the power output. Whenever the overall power output falls below the required level, the condensing turbine is brought into operation to make good the shortfall.

The main disadvantage of a dual condensing/back pressure turbine system as against a pass-out turbine is that units have to be started and stopped frequently, whenever there is a change in the heat/power ratio. There are two variants to the system:

1 Steam from the boiler is fed into the high pressure end of the back pressure turbine. When there is a large demand for process steam, the condensing turbine is fed directly with steam from the boiler. As demand for process steam falls, some of it is upgraded by boiler steam. It is also fed into the high pressure end of the condensing turbine.

2 The two turbines are completely separate from each other. Each is connected to the steam boilers. If an increase in process steam is needed, this is supplied directly from the boiler via a reducing valve. This system is somewhat more costly than the system described in (1) but has improved flexibility during spring and autumn operation.

f Miscellaneous turbine systems

Topping turbines are special, very high pressure steam turbines, which can be run in conjunction with existing back pressure plant. Steam is fed from the low pressure end of these topping turbines into the high pressure end of the existing back pressure turbine. In this way it is possible to cover sudden peaks of power demand without the need to purchase any external electricity.

Diesel plant back pressure turbines use the steam from the diesel plant to drive a turbine, obtaining in the process more electricity and a rather lower level of process heat than is produced by the diesel or gas engine alone.

Double pass-out condensing turbines are employed when steam is needed at more than one pressure. This can be achieved by providing a combined high pressure turbine set and low pressure turbine set. High pressure process steam is taken from the end of the high pressure turbine, while low pressure process steam can be taken from the body of the low pressure set.

Pass-out back pressure turbines are pieces of equipment where the basic process steam is obtained as is usual from the exhaust end of the turbine set. There is, however, an intermediate take-off for process steam, which has to be at a higher pressure than that at the outlet of the back pressure unit.

OPEN CYCLE GAS TURBINES

A gas turbine is basically a machine in which air is taken from the atmosphere, compressed, heated at constant pressure by the combustion of a fuel in air, and allowed to expand once more in a turbine.

Like all heat engines, the efficiency of a gas turbine is governed

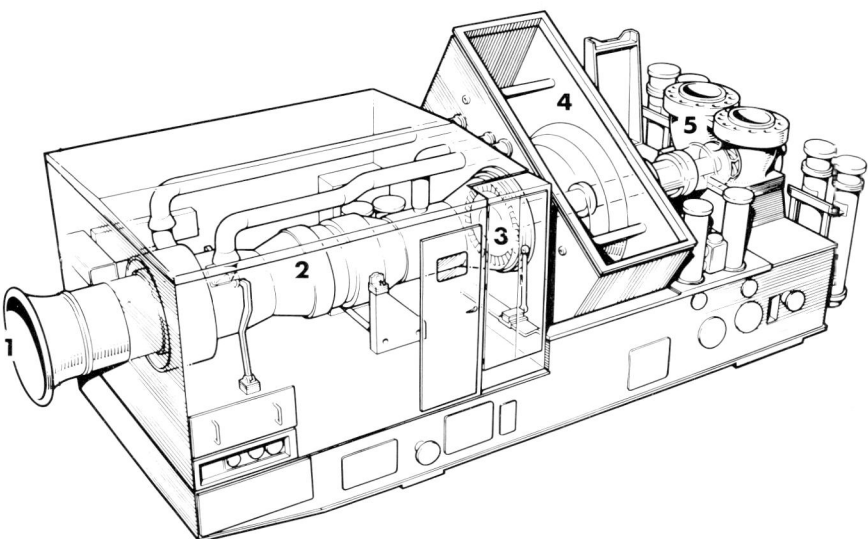

3.3 GEC gas turbine unit
   1 Gas generator air intake
   2 Gas generator

3 Power turbine
4 Power turbine exhaust
5 High pressure pump

mainly by the temperature of the hottest part of the cycle. This in its
turn, depends upon the nature of the materials of which the turbine is
constructed. Modern blading materials enable temperatures of over
800°C to be sustained for long durations. The main use for gas turbines
is in aircraft, but many units are today employed in industry.

Gas turbines have a relatively low capital cost and are also inherently
more reliable than competing prime mover plant. Their main drawback
is the high cost of fuel needed for driving them.

Gas turbines operate at shaft power efficiencies which vary between
about 20 per cent for very simple designs to over 40 per cent for more
complex plant involving intercooling between several stages. From the
total energy point of view open cycle gas turbines have the advantage
that virtually all the thermal energy in the fuel which is not converted
into shaft power is available for recovery.

The steam/gas cycle
The boiler of a conventional steam turbine can use large quantities of
preheated air from the gas turbine exhaust. This still contains a quite
adequate quantity of oxygen to permit further combustion of the fuel.
In the simplest form, the gas turbine discharges directly into the boiler
air intake, all its waste heat being used. The only way in which the
operation of the gas turbine suffers is that a back pressure is imposed
upon it. A variant of such a technique is to abstract the heat in the

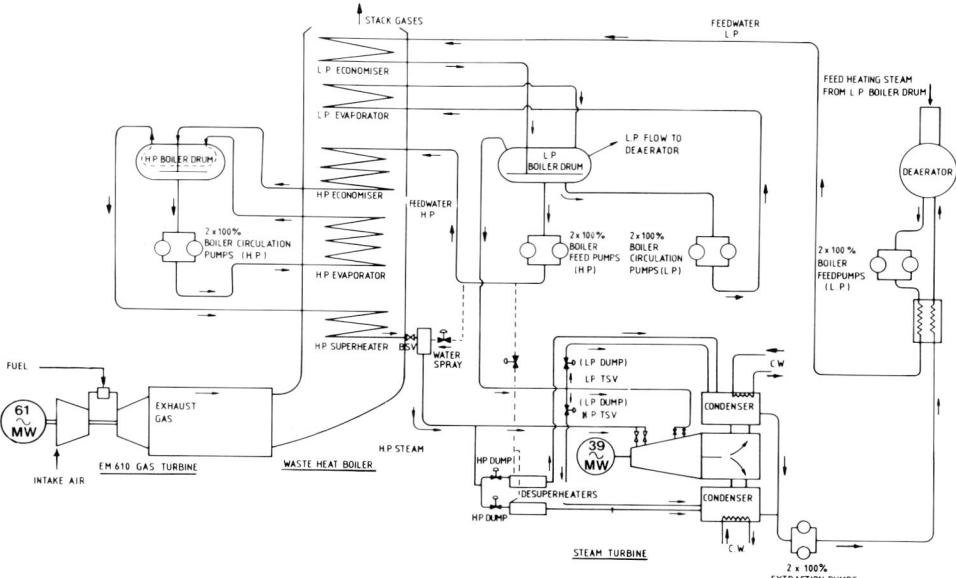

3.4 Total energy combined cycle system producing 100 MW, using GEC EM 610 gas turbine as prime mover. Cycle efficiency 45% (based on LCV).

exhaust gases from the gas turbine by a series of heat exchangers. The following are systems of using waste heat from the exhaust of gas turbines in conjunction with steam turbines, which have been tried:

a Waste heat from the gas turbine is used together with some direct heat to produce high pressure steam. This is fed into a back pressure turbine, which provides extra power and also some useful low level space and process heat. Steam turbines cannot accept quick starts and stops like a gas turbine. It is impossible to have such a steam turbine on the same shaft as the gas turbine unless gearing is employed, as the gas turbine runs at a higher speed.

b Steam can be raised to a pressure rather higher than the maximum pressure of the gas turbine cycle by a combination of exhaust heat and prime fuel combustion. Steam has about half the density but twice the specific heat of air at the same pressure. Provided the water is of high purity this is a good way of utilizing the waste heat of the exhaust to enhance the efficiency of the gas turbine.

c The exhaust gases from the gas turbine are passed through a waste heat boiler to produce steam at 7—15 bars. This is then employed in a standard small condensing turbine to give additional shaft power. In this way it is possible to obtain an overall useful power efficiency of more than 40 per cent with a maximum cycle temperature of about 1150 K. As the steam turbine machinery takes longer to put into operation than the gas turbine plant, it is necessary to have some kind of dis-

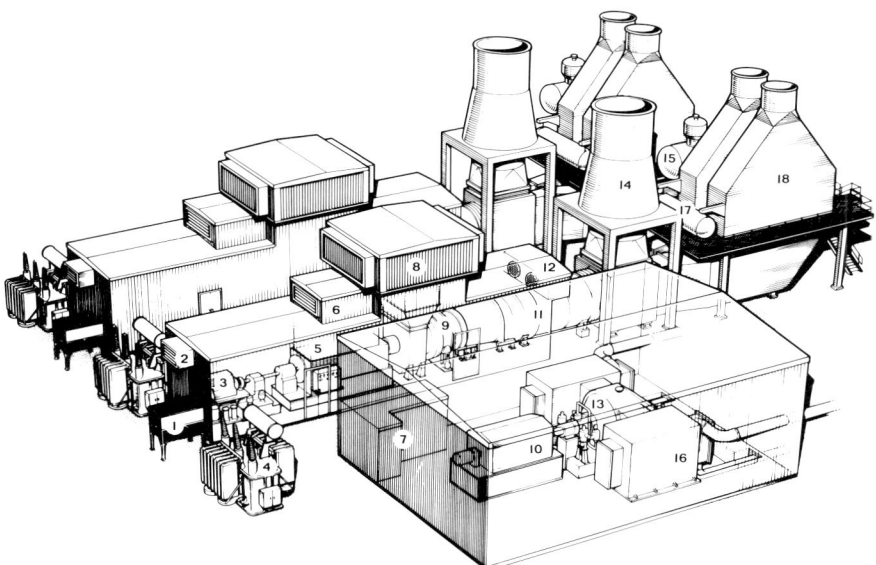

3.5 Layout of GEC gas turbine CHP system

1 Lubricating oil cooler
2 Generator cooling air exhaust
3 Gas turbine starter and
  auxiliaries package
4 Generator transformer and
  coolers
5 Gas turbine generator
6 Gas turbine enclosure
  ventilation inlet
7 Control, switchgear and battery
  rooms
8 Gas turbine air intake unit
9 GEC EM 610 gas turbine

10 Steam turbine generator
11 Gas turbine exhaust diffuser
  and silencer
12 Gas turbine enclosure ventilation
  exhaust and air cooler
13 GEC single-cylinder multi-stage
  steam turbine
14 Bypass stack for gas turbine
  exhaust
15 Deaerator
16 Condenser
17 Steam drum
18 Heat recovery steam generator

connection device, permitting the gas turbine to turn on its own until the steam turbine is ready to run.

Other methods of using the waste heat from gas turbines

Because the exhaust gases from open cycle gas turbines are very hot, (550—850°C) a number of methods of using this heat in plant which does not involve steam turbines have been developed.

Desalination of seawater

Gas turbines are of particular use in countries which are rich in natural hydrocarbons but poor in water, such as Middle East oil producing nations. The exhaust gases are used with large-scale multi-stage flash seawater evaporators to provide fresh water.

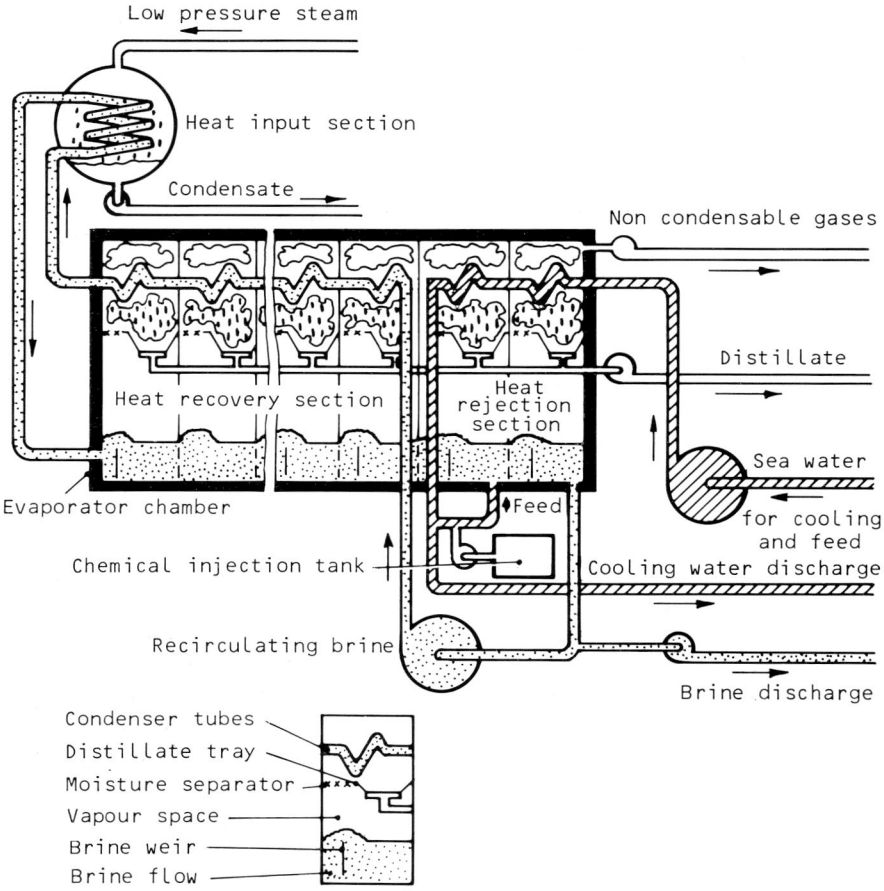

Low pressure steam

Heat input section

Condensate

Non condensable gases

Distillate

Heat recovery section

Heat rejection section

Sea water

Evaporator chamber

for cooling and feed

Chemical injection tank

Cooling water discharge

Recirculating brine

Brine discharge

Feed

Condenser tubes
Distillate tray
Moisture separator
Vapour space
Brine weir
Brine flow

3.6 Use of waste heat for desalination. Courtesy Weir Westgarth

A flash evaporator can consist of twenty to fifty chambers, each operating at a lower pressure than the preceding one. As heated brine flows
from one chamber to the next, some of it flashes off to form water
vapour and more concentrated brine. The water vapour condenses on
the colder condenser tubes and drops as distillate into trays, from which
the fresh water is then led to storage tanks. The brine, as it passes from
chamber to chamber, cools down progressively and is finally pumped
back through the condenser tubes to act as coolant in the condenser section of each chamber. It becomes progressively hotter as it does so. In
consequence, when it reaches the heat input section and before entering
the first flash chamber, its temperature need only be raised a few
degrees to allow the vapour released in the flash chamber to condense
on the tubes. This heat is supplied from the exhaust gases of the gas
turbines.

The efficiency of flash evaporator plants is normally expressed by the following formula:

$$\text{Th. E.} = \frac{(t_c - t_1)\,L}{(t_c - t_f)} \text{ kJ/kg water produced}$$

where Th. E. = kJ of energy used per kg of fresh water

t_c temperature of water entering flash stage in °C

t_1 temperature of water leaving tube system in °C

t_f temperature of water leaving last flash stage in °C

L   effective latent heat of water in kJ/kg

The latent heat of water at 100°C equals approximately 2257 kJ/kg. If one used only single evaporation, this would be the heat needed to obtain 1 kg of pure water.

When using flash distillation, however, one obtains a ratio

$$\frac{(t_e - t_1)}{(t_e - t_f)}$$

which varies between 0.1 and 0.25, depending upon the nature and the design of the flash evaporator concerned.

This means that the heat needed to produce 1 kg of fresh water from brine varies between 230 and 560 kJ. The lower figure is generally only obtainable from very large and complex multiple flash evaporation systems with numerous flash chambers. These are also very expensive. In general, it is uneconomic to use such very expensive plant when the heat used for reheating purposes is, in fact, waste heat from prime movers.

Provision of process steam

Waste heat from prime movers can be converted into process steam which is used by a wide variety of industries. Paper mills use large quantities of such steam but they usually require it at varying pressures and at odd times. One way of overcoming this problem is by the provision of steam accumulators. If water is stored close to its critical temperature (374.15°C) at which the pressure equals 221.2 bars, its latent heat of evaporation can be reclaimed once the pressure is released.

Modern pre-stressed cast iron vessels are capable of storing up to 18 m³ of water at pressures as high as this. Steam is supplied from such stored high temperature water simply by flashing off. Theoretically, if the water is kept at the critical point, the entire quantity of approximately 18 tonnes of steam can be stored in this one vessel without the temperature of the residual hot water dropping at all. In actual fact, such a vessel is probably only able to store about 15 tonnes of steam. The chemical industry uses enormous quantities of process steam for distillation, fractionation, evaporation and other unit operations. Breweries, food processing industries, laundries, hospitals and many others are also ready consumers of process steam which can be provided from waste heat produced by prime movers, either with or without energy storage.

Direct drying
There are a large number of industrial processes where waste heat from prime movers is used for drying purposes. The most common of these are the following:
Brick, tile, ceramic and glass manufacture
Leather and allied trades
Agricultural establishments such as plants used for vegetable drying, haymaking etc
Fishmeal and meatmeal factories

Overall thermal efficiencies of power plus heat production are above 70 per cent in cases like these.

DIESEL AND GAS ENGINES

With modern diesel and gas engines in which both the cooling water and the exhaust gases are at high temperatures, it becomes possible to use the waste heat produced as steam, hot water, or both. When compared with other prime movers, diesel engines have excellent thermal efficiencies. Up to 39 per cent of the calorific value of the fuel can be converted into shaft power. There are three sources from which heat can be recovered, when operating diesel and gas engines. They are:
1 Exhaust gases
2 Jacket water heat extraction
3 Lubricating oil cooler

Exhaust gases
In the case of modern turbocharged four-stroke engines, the mass flow of exhaust gases is roughly 7.9 kg/h per kW of shaft output. A fairly typical 2000 kW diesel engine therefore emits 15.8 t of exhaust gases each hour.

The temperature of these exhaust gases after the turbo-charger is approximately 400°C at full load. Heat can be extracted from the gases by passing them through the tubes of a boiler or water heater, the minimum outlet temperature being approximately 175°C. For steam boilers the exhaust outlet temperature must be approximately 40°C higher than the saturation temperature. This maintains the tube lengths within economic limits and reduces corrosion troubles at low engine loads when exhaust temperatures fall.

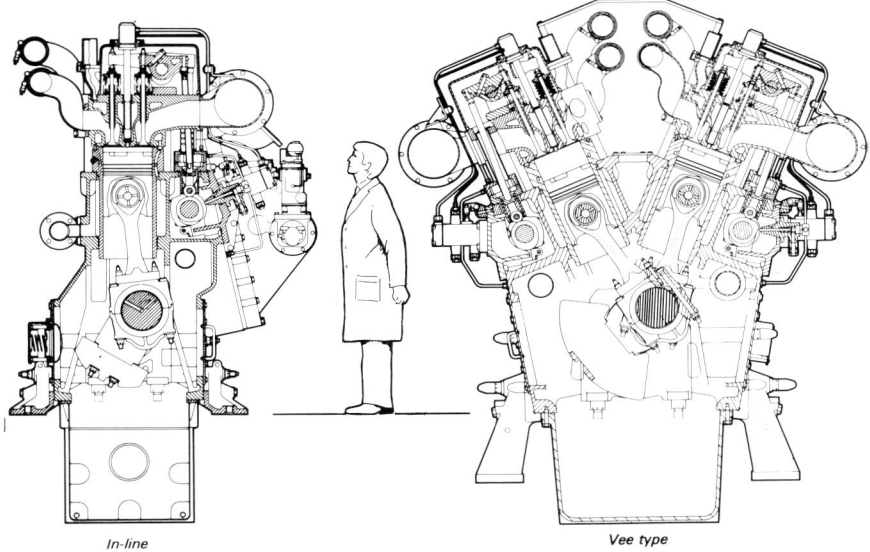

In-line   *Vee type*

*3.7 Two large Mirrlees-Blackstone MB 275 diesel engines suitable for total
energy use*

If the specific heat of the gases is 1.0467 kJ/kg K, the maximum
amount of heat which can be taken off from the exhaust gases is thus:
7.9 × 1.0467 (400 − 175) kJ/kWh = 1860 kJ/h per kW output, of the
prime mover concerned, while it is on full load.
A typical 2000 kW shaft power diesel engine is thus able to yield

$$\frac{1860 \times 2000}{3600} \text{ kW} = 1033 \text{ kW of usable heat from its exhaust alone.}$$

Types of exhaust gas boilers

There are two types of exhaust gas boilers or water heaters, namely vertical and horizontal ones. Both consist of gas inlet and outlet boxes, and
a calandria of tubes, which is immersed in water. In any specific station
containing three or less boilers, a feed pump would supply each boiler.
Standby pumps are also supplied. These pumps are started and stopped by float-level controllers in the boilers, which are also fitted with
alarms. In a station which contains a large number of boilers it is a better arrangement to have one or two continuously running feed pumps.
These supply a pressurized feed main from which water is admitted to
the individual boilers by feed controllers. Two engines can feed a common boiler of the double inlet, double outlet type in which the gas sections are entirely separate, but the water section is common. It is not
recommended that the exhaust gases of two or more engines are mixed,

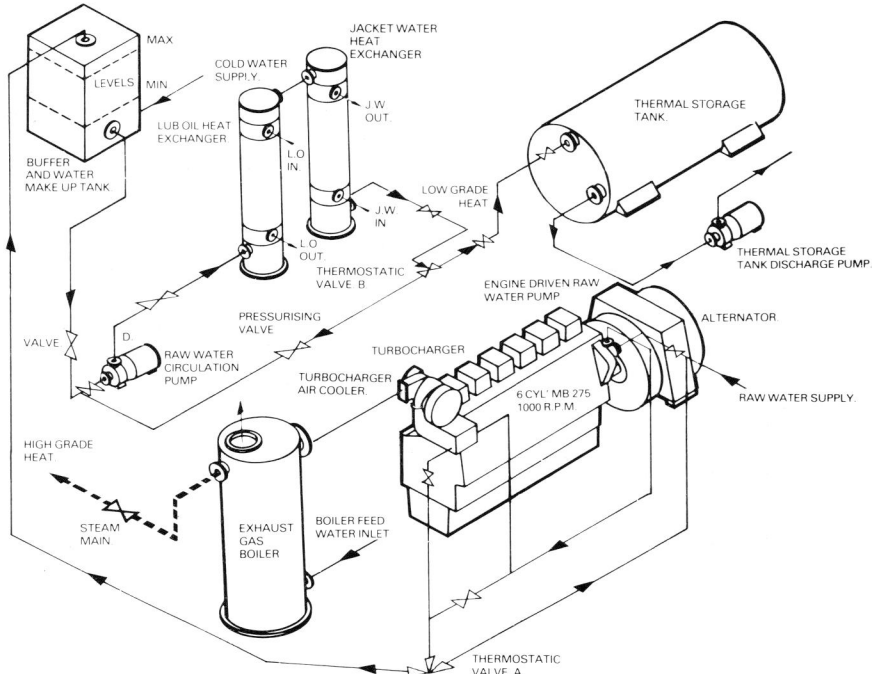

3.8 Operation of Mirrlees-Blackstone 510 kW diesel engine in total energy system

as this can lead to difficulties with back pressures and also with the necessary gas isolating valves.

The exhaust gas outlet temperature from a boiler increases with rising steam pressure. In consequence, the quantity of steam produced falls. It is therefore inadvisable to use steam pressures in excess of 7 bars for heating loads and 17 bars for power loads.

A further disadvantage of high steam pressures is that steam production drops rapidly when the engine load is reduced, because the exhaust gas temperature falls.

The following table gives yields of steam with a typical 2000 kW shaft power diesel engine.

TABLE 3.2 STEAM YIELD FROM A 2000 kW DIESEL ENGINE

Pressure of steam in bars gauge	kg of steam per hour
1.00	1540
3.40	1450
7.00	1260
17.00	1000

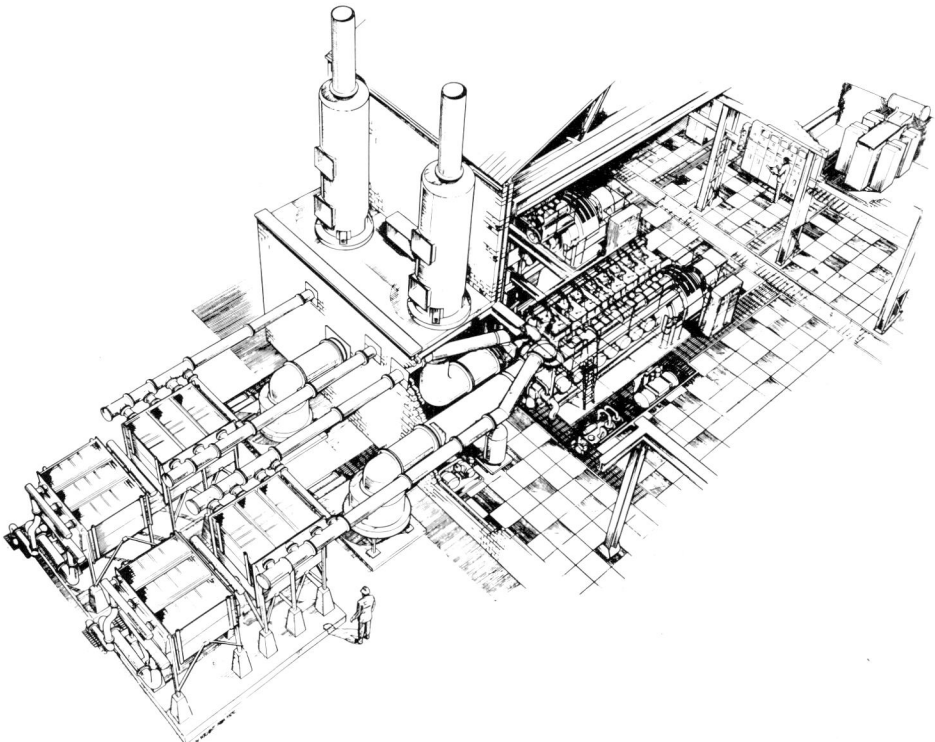

3.9 Layout of a total energy plant using Mirrlees-Blackstone K8 diesel engines as prime movers

If an exhaust gas water heater is installed, the maximum water temperature is fixed by the system's pressure and is usually at around 93°C. If the heat input to the water heater is constant, a given water flow rate produces a fixed temperature rise. If, however, there is a variation of heat input, a constant flow rate results in varying temperatures. To maintain a steady temperature it is then necessary to provide auxiliary heating.

Jacket water heat extraction

Auxiliary heat exchanger cooling

To provide secondary water at as high a temperature as possible the engine jacket water outlet is set at 82°C, which gives secondary water from an auxiliary heat exchanger at about 77°C. When no hot water is needed, jacket water is passed through a main heat exchanger which is fed by raw water from the cooling system. A thermostatic diverter valve controlled by the jacket water inlet temperature to the engine valve must be fitted in the raw water circuit. This is necessary to prevent

3.10 A GEC waste heat boiler unit
A Incoming flow from gas turbine
B Gas entering boiler tube banks
C Gas exhaust through boiler stack

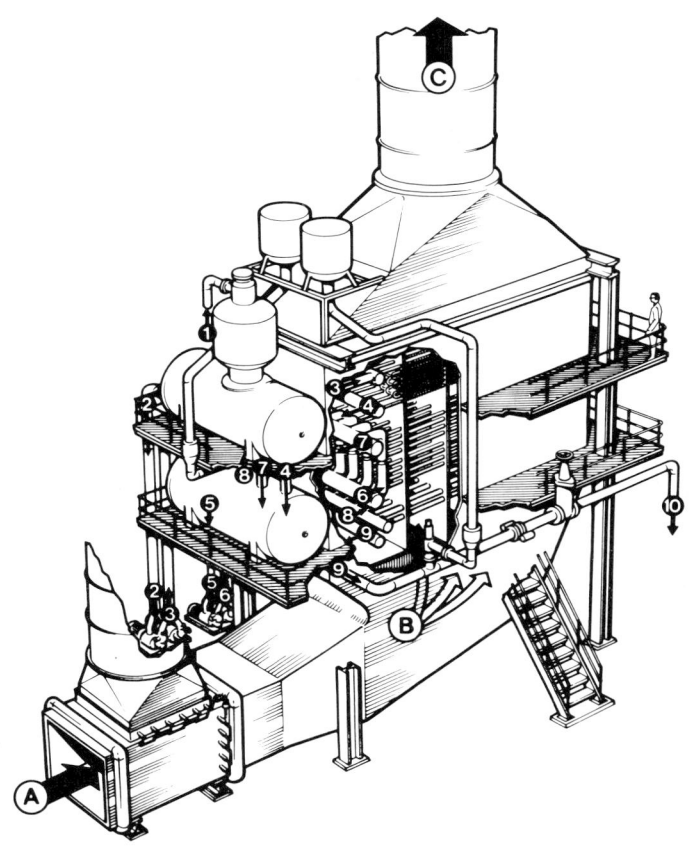

1 Incoming water to water deaerator
2 From deaerator to feed water pumps
3 From feed water pumps to economizer inlet header
4 From economizer outlet header to drum
5 From drum to circulator pumps
6 From circulator pumps to evaporator inlet header
7 From evaporator outlet header to steam drum
8 From drum to superheater inlet header
9 From superheater outlet header
10 Superheated steam supply to steam turbine

under- or over-cooling due to variations in the secondary water flow rate, water temperature or engine load.

Latent heat cooling

The jacket water is pressurized up to about 2 bars absolute, so that its water outlet temperature is around 120°C. When the hot jacket water passes to a steam separator, which is at atmospheric pressure, steam is flashed off. The water itself is cooled to 99°C by the abstraction of latent heat, at which temperature it is returned to the cooling jacket of the engine. It is necessary to condense the steam to obtain a heat load, and to return the condensate to the jacket water. Alternatively, extra cooling water is supplied, equal in quantity to the mass of steam extracted.

Advantages of latent heat cooling are:

1 The engine jackets are maintained at a high temperature. The high cost of distillate fuel oils has meant that an increasing number of diesel engines use residual fuel oil rich in sulphur. During combustion this sulphur burns to sulphur dioxide, which oxidizes to sulphur trioxide. If this is permitted to dissolve in water, harmful sulphuric acid is formed. As the dewpoint of the products of combustion of a diesel engine is about 90°C, the maintenance of a high temperature prevents harmful condensation from taking place.
2 The engine jackets are maintained at a constant temperature, and the rise in water temperature across the engine is slight. Unless a thermostatic water division valve is used, a conventionally cooled engine suffers from fluctuating jacket water temperatures.
3 The lubricating oil remains free from water, acid and sludge.
4 There is an improvement in fuel consumption.
5 Steam at 1 bar gauge is produced, which can be used for space heating or process heating with most installations.

Using latent heat cooling, one can reckon on obtaining about 2800 kJ of heat each hour per kW shaft power or 5600 MJ per hour for a 2000 kW diesel plant.

Lubricating oil coolers

A full load temperature of 70°C can be maintained safely for the lubricating oil which leaves an engine. In consequence, secondary water at a temperature of about 57°C can be obtained at the oil cooler outlet. Engines which are fitted with oil cooled pistons reject approximately 420 kJ per hour for each kilowatt engine output. A typical 2000 kW engine therefore delivers 840 MJ of heat each hour from its oil coolers in the form of low grade heat. However, because this heat is at such a low temperature grading, it is not usually extracted unless one wishes to aim at maximum heat recovery with a given engine. Its comparatively low temperature makes it suitable only for such purposes as the pre-heating of very cold water or for use in an underfloor heating system. Auxiliary oil coolers tend to be rather expensive.

The main uses for waste heat from diesel and gas engines are the following:

Space heating. In summer it is necessary that some other use be found for this waste heat. It is possible to employ well-insulated underground heat storage systems. These accept the heat during periods when space heating loads are low, to augment the heating capacity during the winter. Some of the waste heat can be used to run air conditioning plant, but much is of too low a grade to do this efficiently.

Process work. Recovered heat can often be used to full advantage within the limits of water temperature or steam pressure obtainable. Closed circuit operation should be used to obviate contamination.

Additional power. When diesel engines of the size range of 15—20 MW are used, steam conditions from exhaust boilers approaching 340°C and 18 bars gauge can often be obtained. Such steam can usefully drive other turbines. The shaft power output of such sets is unlikely to rise above 5–7 per cent of that of the diesel engine. In most cases such a system would not be cost-effective.

Main boiler feed heating. There is a commonly accepted rule that for every 6°C by which the feed water temperature of a boiler is raised, there is a 1 per cent saving in prime fuel costs.

The extent to which feed water can be heated by the use of waste heat from diesel engines depends upon its nature. Engine jacket water can be used up to a temperature of 77°C, while 1 bar gauge steam can preheat the feed water up to 113°C, representing boiler fuel supply savings of 5.5 and 11.4 per cent, respectively.

Heavy fuel heating. Heavy fuel must be pre-heated in order to lower its viscosity. For example, 3500 seconds Redwood No 1 oil at 38°C must be heated to 113°C to use in a diesel engine.

Refrigeration. Waste heat from a diesel plant is usually of too low an order to employ it in a compression plant. It is adequate for employment with a lithium bromide absorption plant.

Waste heat from diesel and gas engines can also be used to operate multiple flash desalination plants and sewage plants.

Free piston gas engines
In the case of a free piston generating plant, there are two loose pistons which take the place of the compressor and combustion chamber of a gas turbine. They can convert many different types of fuels into hot

gases to drive an associated gas turbine. A combined free piston generating plant and gas turbine together have about the same efficiency as a
diesel engine, which is higher than that of a straight run gas turbine.
Yet in terms of capital costs and weight, the system lies closer to the gas
turbine than to the diesel engine. Repair costs are high and units below
25 MW in size are not particularly economic. Gas turbines fed by free
piston gas generators operate at a lower temperature than straight gas
turbines, namely around 500°C instead of 700°C as is normal. This
enables poorer grades of fuels to be used. Fuels up to 5 per cent sulphur, and fuels adulterated with sodium and vanadium which would be
impossible to use in normal gas turbines, can be employed with free pis-
ton/gas turbine plants.

Free piston gas engines use water-cooled cylinders and compression
heads, while pistons are oil-cooled. The heat removed by the cooling
system is about 14 per cent of the heat content of the fuel, and can be
used for low grade heat. Small free piston/gas turbine plants run at
efficiencies of up to 34 per cent while larger units can reach 38 per cent.

The stirling engine

In the stirling engine thermal energy is converted into mechanical
energy by first compressing an enclosed quantity of working gas at a
comparatively low temperature, heating it and allowing it to expand at a
higher temperature (about 700°C in practice). The piston which closes
the gas filled working space needs to do less work to compress the gas at
a lower temperature than the amount of work which is performed at the
high temperature. There is therefore a net gain of mechanical work
from this engine for the expenditure of some thermal energy.

N. V. Philips of Holland were carrying out research on stirling
engines in collaboration with the Ford Motor Corporation. This work
has now been terminated.

PRACTICAL EXAMPLES OF TOTAL ENERGY INSTALLATIONS

1 A Mirrlees Blackstone MB275 1600 kW diesel engine was coupled to
a Spanner waste heat boiler to raise up to 860 kg/h of steam at 7.5 bars
from the diesel engine exhaust gas at 365°C. The plant was installed at
the Harp Lager brewery at Dundalk and cost about £½ million in 1981.

High grade heat is obtained from the exhaust gases in the form of
steam, while low grade heat is recovered from the engine jacket and lubricating oil cooling water to provide hot water via Serk heat exchangers.
A 255 m³ hot water storage facility comprising three tanks provides the
necessary thermal buffer to absorb the variations in the brewery's hot
water requirements.

The following combined heat and power energy breakdown is given:

Fuel used: 329.3 litres of oil with CV 41 MJ/litre equals input of 13 500 MJ for one hour's operation

Output:

Stack and other losses	2 450 MJ	=	18.2%
Electricity	5 400 MJ	=	40.0%
Steam at 7.5 bars	2 510 MJ	=	18.6%
Hot water at 68°C	3 140 MJ	=	23.2%
	13 500 MJ	=	100.0%

Efficiency at total energy plant = 81.8 per cent

One of the main reasons why the Harp brewery scheme has been a success was that the various loads for the different energy elements—electricity, steam and hot water—all exist in a fairly convenient balance.

2 At Ras Abu Fontas in the state of Qatar a system of 12 large gas turbines, each with an output of 50 MW, was installed. Their exhaust is passed through waste heat recovery boilers which supply low pressure steam to twelve multi-stage flash distillers. To achieve the necessary flexibility of operation, the required steam flow to sustain full water production can be obtained by auxiliary fuel firing, whenever the gas turbines are operating at below 80 per cent of their rated capacity.

Each of the Weir Westgarth multi-stage flash distillers is able to supply 22 500 m^3 of fresh water per day. The average ambient air temperature at which the Ras Abu Fontas plant is being operated is 30°C. At this temperature the net power output equals 26.3 per cent of the calorific value of the feed fuel, and the overall operating efficiency may be expressed as:

$$\frac{\text{electricity} + \text{heat utilized}}{\text{heat content of fuel}} = 64.0 \text{ per cent}$$

The turbines can burn either crude oil, associated oil-gas or non-associated oil-gas. Combined gas turbine plants and multiple flash desalination plants which use steam raised from the exhaust are considered to have many advantages under the conditions which exist in Middle Eastern oil states.

3 Another brewery, this time Mitchells and Butlers of Birmingham, recently installed an SSK steam turbine made by W. H. Allen of Bedford. At maximum steam flow of 31.8 tonnes per hour at 20.5 bars exhausting to 4 bars, the unit generates 1 146 kW.

Average process steam demand is 17 036 kg per hour, the turbine exhausting at 95.5 per cent dryness fraction. It produces an average of 582 kW over 7900 hours of annual operation. With a boiler efficiency of 73 per cent it was found that the steam costs amount to 0.225 p per 1269.47 kJ. The enthalpy change as the steam passed through the turbine amounted to 148.63 kJ/kg.

It is therefore possible to evaluate the annual cost of generating 582 kW as:

Boiler steam flow (kg/h) $\times$ enthalpy change (kJ/kg) $\times$ steam cost ($£$/kJ) $\times$ hours run (h)

or: $£\ \dfrac{17\,036 \times 148.63 \times 0.225 \times 7\,900}{1269.47 \times 100} = £35\,453$

In contrast, the Midland Electricity Board charged £3.18 p/kW h for electricity supplied, so that the annual cost of the same 582 kW from public supply would have come to:

$£\ \dfrac{582 \times 3.18 \times 7\,900}{100} = £146\,210$

making an annual saving of approximately £110 760 for a capital investment of around £150 000.

Even assessing extra costs incurred in operating the total energy system, a payback period of approximately two years is obtained.

Checklist

1 Electricity generation is a low-efficiency conversion of primary energy.

2 Total energy operation uses the low grade waste heat produced during power generation.

3 There are numerous advantages and disadvantages in operating total energy plants as against purchasing power and using a separate boiler for works heat needs.

4 There are three basic types of steam turbine: condensing turbines, back pressure turbines and pass-out turbines.

5 When gas turbines are used, exhaust heat should be duly utilized, being extracted either through a steam/gas cycle or by waste heat boilers.

6 Open cycle gas turbines are particularly useful in conjunction with desalination equipment in hot and arid countries.

7 Systems using diesel and gas engines have good overall efficiency when waste heat is extracted from the exhaust gases, the jacket water and the lubricating oil cooler.

8 Two less usual sources of total energy are of free piston gas engines and stirling engines.

SELECTION CHART: TOTAL ENERGY EQUIPMENT

Type of equipment	Advantages	Disadvantages	Appropriate applications
Steam turbine systems of all types	Cheap fuel Easy control to vary heat/power ratio according to specific demand	Only feasible for very large systems Power sacrifice when heat is extracted	Urban district heating systems, very large industrial undertakings
Open cycle gas turbines	Little change in efficiency when operating on total energy High heat grading	Fairly costly fuel Only feasible in sizes larger than would be possible with gas and diesel engines	Large and medium size factories, municipal complexes, farms (crop drying), desalination of seawater
Closed cycle gas turbines, stirling engines and free piston gas engines	Cheap fuel Good efficiency with small sets	Expensive to make and to maintain	Not as yet properly perfected Still only in prototype stage
Gas and diesel engines	Good operational characteristics with small sets	Costly fuel High purchase and maintenance costs Poor efficiency of power generation	Small factories, farms

References and further reading

1 R. M. E. Diamant, *Total Energy*, Pergamon Press, Oxford 1970.
2 G. H. Platt, 'Waste heat recovery—the steam turbine in combined cycle operation', *APE Engineering*, 27, 1981, p 2–7.
3 T. Gibson, 'Fluidized bed CHP plant powers UK dyeing factory', *Modern power systems*, November 1981, p 44–46.
4 *APE Engineering*, a regular journal published by NEI-A.P.E. Ltd, Bedford, UK.
5 Technical information issued by Stal-Laval AB of Finspong, Sweden.
6 Technical information issued by Mirrlees Blackstone Ltd, Stockport, Cheshire, UK.

SELECTED COMPANIES INVOLVED IN MANUFACTURE OF PRODUCTS

UK and Europe

NEI-A.P.E. Ltd, PO Box 43, Bedford MK40 4JB.
ASEA AB Vasteras, Sweden S–721 04.
Beverley Chemical Engineering Ltd, Billingshurst, West Sussex RH14 9SA.
E. J. Bowman Ltd, Chester Street, Birmingham B6 4AP.

Born Heaters Ltd, Europa House, Goldstone Villas, Hove, East Sussex BN3 3RZ.
Dale Electric Ltd, Electricity Buildings, Filey, Yorks YO14 9PJ.
Earby Light Engineers Ltd, Datcliffe Works, Kelbrook, Earby, Colne, Lancs BB8 6TW.
GEC Gas Turbines Ltd, Whetstone, Leicester, LE8 3LH.
Howden Engineering Ltd, Clipstone House, Hospital Road, Hounslow, Middx TW3 3HT.
IEI Northern Ltd, Welfield House, Victoria Road, Morley, Leeds, LS27 7PA.
A. Johnson Ltd, Aldwych House, Aldwych, London WC2B 4EL.
Lucas Aerospace Ltd, Monkspath, Shirley, Solihull B90 2JJ.
Mirrlees Blackstone Ltd, Hazel Grove, Stockport, Ches SK7 5AH.
Noel Penny Turbines Ltd, Siskin Drive, Toll Bar End, Coventry CV3 4FE.
NV Philips, PO Box 523 5600, AM Eindhoven, Holland.
Rolls-Royce Ltd, Ansty, Coventry CV7 9JR.
Siemens GmbH, Erlangen, West Germany.
Stal Laval Turbin AB, S 61220 Finspong, Sweden.
Sulzer Bros, Farnborough, Hants.
Walton Engineering Ltd, 50 Pall Mall, London SW1Y 5JR.
Wells Spiral Tubes Ltd, Prospect Works, Airedale Road, Keighley, West Yorkshire BD21 4LW.

United States
Air Research Manuf. Co., 2525 T W 190th Street, Torrance, CAL.
Carrier Corporation, Carrier Tower, PO Box 4800, Syracuse, NY.
Caterpillar Tractor Co., 100 NE Adams Street, Peoria, IL 61629.
Cummings Power Inc., 5100 E 58th Avenue, Commerce City, COL 80022.
Diesel Engines Systems Inc., 77 Mark Drive, Suite 11, San Rafael, CAL 94903.
Ford Power Products Inc., 300 Renaissance Center, PO Box 43328 T, Detroit, MI 48243.
Gas Turbine Corporation, 4430T Director Drive, San Antonio, TX.
General Electric Co., 1 River Road, Schenectady, NY.
Lister Diesel Inc., 555 East 56 Hiway, Olatte, KS 66061.
Omega Industries Inc., 99 Jericho Turnpike, Jericho, NY 11753.
Onan Corporation, US Power Products Division, 1400 73rd Avenue NE, Minneapolis, MN 55432.
Rotoflo Corporation 2235T Carnelina Avenue, Los Angeles, CAL.
Skinner Engine Co., PO Box 1149, Erie, PA 16512.
Trane Co., Thomas Creek Road, La Crosse, WI 54601.
Western Engine Co., 750 RT53, Itasca, IL 50143.
Westinghouse Corporation, PO Box 158, Madison, PA 15663.

4 Fuel cells

A torch battery is, theoretically speaking, an example of a fuel cell. An internal rod of carbon surrounded by a manganese dioxide depolariser is the cathode of the cell. On the outside, there is an envelope of zinc metal. A chemical reaction takes place between the zinc and the electrolyte, which is a paste of ammonium chloride in starch.

$$Zn \rightarrow Zn^{2+} + 2e \qquad \text{(constituting the anode reaction)}$$
$$2NH_4^+ + 2H_2O \rightarrow 2NH_4OH + 2H^+ \qquad \text{(in the electrolyte)}$$
$$Zn^{2+} + 2Cl^- \rightarrow ZnCl_2 \qquad \text{(final anode product)}$$
$$2H^+ + 2MnO_2 \rightarrow Mn_2O_3 + H_2O - 2e \quad \text{(cathode depolarizer reaction)}$$

An emf of 1.5 v is applied between anode and cathode.

It can be seen that during the torch battery reaction zinc at the anode has been converted to zinc chloride, while at the cathode the manganese dioxide has been changed permanently to Mn_2O_3. Electrical energy has been given out and the fuel components, zinc and manganese dioxide, have been converted into waste materials.

Thermodynamically we can express what happens by writing that:
$$\Delta G = -nFE$$
where ΔG is Gibbs free energy in J/mole

        n is the number of electrons liberated per atom of fuel burned (2 in the case of zinc)

        F is Faraday's constant = 96 500 As/mole

        E is the effective emf of the cell concerned in volts

Thermodynamic considerations

For the average practical torch battery which, as already mentioned, gives out 1.5 v

ΔG works out at $-2 \times 96\,500 \times 1.5$ kJ/mole $= -289.5$ kJ/mole (65.37 grammes) of zinc.

The amount of electrical energy which is produced by the conversion of a mole of zinc into Zn^{2+} ions, or as in this case zinc chloride, is thus equal to 289.5 kJ or 80.42kWh. Unlike the generation of electrical energy by the burning of a carbonaceous fuel, no (or very little) heat is produced. There is no conversion of chemical energy into heat energy, thereby neutralizing the effect of Carnot's equation, which has such a devastating effect upon the actual electricity production efficiency of normal heat engines.

Gibbs free energy change ΔG is related to the enthalpy change ΔH by the following equation:

$$\Delta G = \Delta H - T\Delta S$$

Where ΔH is the enthalpy change in J/mole

ΔS is the entropy change in J/mole K

T is the absolute temperature at which the reaction takes place, in kelvins.

Theoretically, at any rate, there should be no entropy change with an ideal conversion from fuel into electric power and therefore

$$\Delta H = \Delta G$$

This means that the work done by the fuel cell ΔG represents the entire change in enthalpy ΔH between the fuel and the waste products formed. In practice, however, this does not really happen. There are several irreversible processes which cause a considerable difference between ΔG and ΔH and the thermodynamic efficiency of the process is reduced to well below 100 per cent.

Practical fuel cells

A torch battery only works for a time and has then to be thrown away because both the fuel and the waste products are solid and cannot be replaced. Practical fuel cells have to use fluids, either gaseous or liquid substances, for both fuel and waste products, so that they can be easily fed in and out of the system.

The most important fuel is hydrogen, which is burned by means of oxygen to water, which in its turn is eliminated as a waste product. Alternative fuels are a wide variety of hydrocarbons, alcohols, and such substances as ammonia and hydrazine.

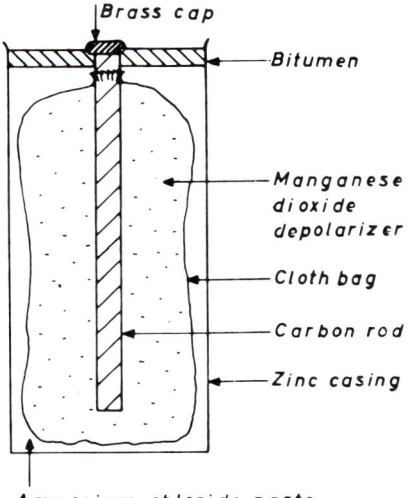

Brass cap

Bitumen

Manganese dioxide depolarizer

Cloth bag

Carbon rod

Zinc casing

Ammonium chloride paste

4.1 The torch battery or Leclanché cell, and how it works

Application of fuel cells

As already indicated, fuel cells are very efficient at converting fuel into electricity. This is because the conversion is direct and does not pass via the heat engine as happens with steam turbines, diesel engines and petrol engines. The efficiency of fuel conversion into electric energy ranges from about 25 per cent at a power output of 1 kW to over 50 per cent with very large installations.

In contrast, small-scale electric motors or petrol-driven plants have

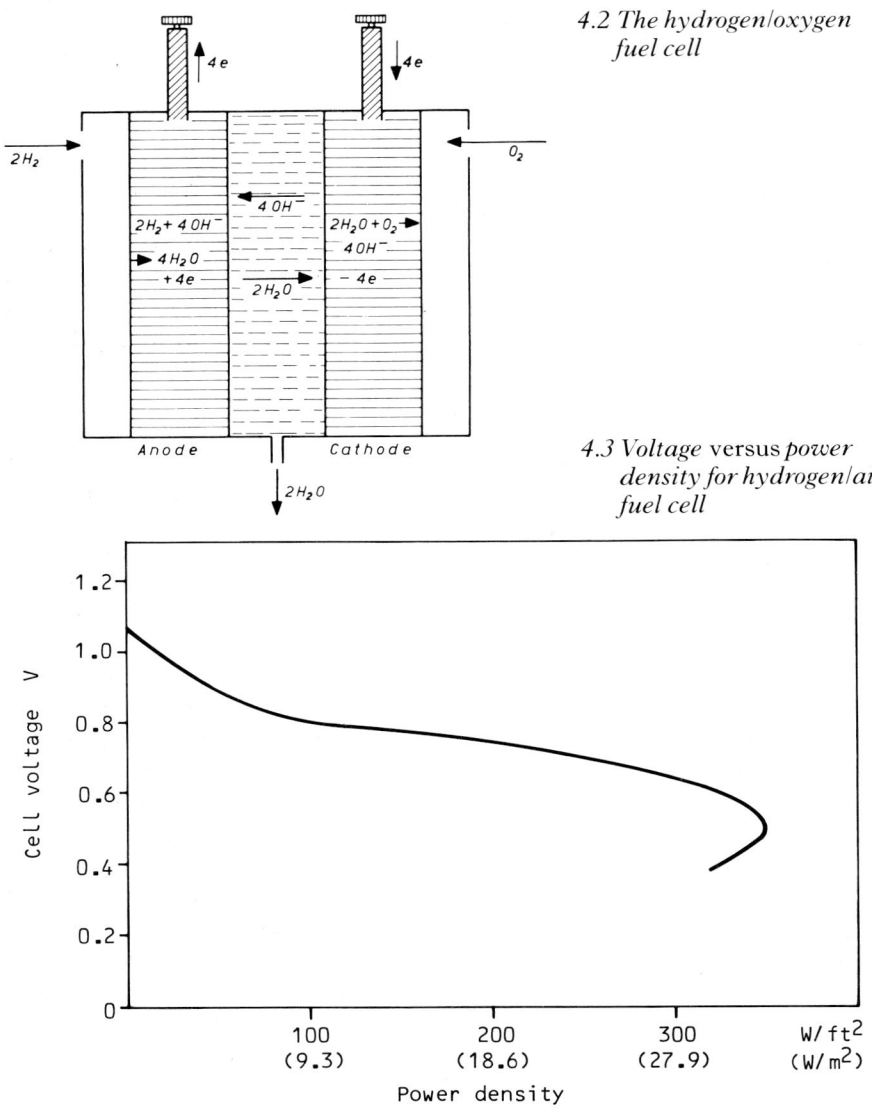

4.2 The hydrogen/oxygen fuel cell

4.3 Voltage versus power density for hydrogen/air fuel cell

thermal efficiencies of only about 10 per cent, while even the largest and most elaborate steam turbine systems do not yield more than around 35 per cent of the net calorific value of the fuel used as electric energy. Once loading of a large steam power station is reduced, the thermal efficiency falls drastically. This is not the case with fuel cells.

Fuel cells are at their most valuable when used in small installations because under such circumstances there is the biggest difference in efficiency between them and conventional power generation equipment. Local generation of electricity, avoiding long transmission lines, becomes feasible. Fuel cells have no moving parts and are therefore desirable under conditions where noise has to be avoided. Low pollution is another advantage, as against equipment in which fuels are burned to provide heat.

The most promising application to date are the following:
a For spacecraft where the low weight and high reliability of hydrogen/oxygen fuel cells have been found to be of enormous value. The Gemini project used 1 kW fuel cells constructed by the GEC company, while the Apollo fuel cells, built by Pratt and Whitney, were capable of 2 kW output each. Both types had an excellent power to weight ratio. The sole waste product was water, which was suitable for drinking by the astronauts.
b Fuel cells have considerable application in propelling non-nuclear submarines, because no harmful products are produced by combustion. Nuclear reactors are only suitable for larger vessels. The most successful fuel cells so far used are the Alsthom hydrazine/hydrogen peroxide cells, capable of yielding up to 30 kW per cell.
c Fuel cells have been employed, either on their own or with batteries, for electrically powered vehicles. They are particularly useful for fork lift trucks, agricultural machinery, construction machinery and milk floats. Fuel cells overcome the problems of short range and excessive battery weight which usually beset electric vehicles. At the same time, they offer the advantages of easy stop/start operation and pollution-free working which are a feature of electric vehicles. Hybrid power units in which batteries supply peak energy while fuel cells recharge the batteries continuously seem the most promising. Several experiments have been carried out to investigate the suitability of such systems for military purposes and for private cars. It seems at present rather unlikely that hybrid power systems will be an economic proposition in the short term for motor car propulsion, as costs are still far too high. Economic criteria are however of lesser importance for military use.
d Fuel cells have been used widely for powering remote relay systems and military communications systems. The US army uses a 60 watt Monsanto fuel cell which burns hydrazine and weighs only 5½ kg and can operate for 24 hours on one kg of fuel.

e Much experimental and development work is now proceeding to design commercial fuel cell power plants, able to generate electricity from fossil fuel. Attempts are being made to develop equipment which is more efficient at power generation than heat engines, and which produces neither pollutants nor noise. The main problem to be overcome is that the capital cost of a fuel cell plant per unit power output is many times higher than that of competing heat engine systems.

f Fuel cells have considerable promise in applications where their mode of operation is particularly advantageous. The following are examples:
Operation of artificial kidney machines.
Environmental control systems such as air cleaners, dehumidifiers and water purification plants.
All kinds of hospital purposes.
In miniature form for artificial heart power supply.

Costs of fuel cell systems

While for space and military applications the cost of a fuel cell system is of less importance than its performance, this is not so with normal commercial applications, where cost is the most vital factor involved.

The most expensive fuel cells were those used for space exploration purposes, costing up to $400 000 per kW output. Fuel cells for military purposes cost between $10 000 and $40 000 per kW, but it will be necessary to lower the cost of stationary hydrocarbon fuel cells developing about 15 kW to below $200 per kW. At the time of writing the best achieved is a cost figure of $800 per kW. It is estimated that even today it would be possible to produce fuel cell power plants on a reasonable series production run, for approximately $350 per kW.

This would indicate that as fossil fuel prices rise and manufacturing techniques for fuel cells improve, we may soon reach the balance point at which fuel cells become viable.

TYPES OF FUEL CELL

Direct fuel cells
These use the fuel directly to produce electrical energy. By far the most common is the hydrogen/oxygen fuel cell, which is available in low temperature, intermediate temperature and high temperature forms.

The following alternative fuels can be used instead of hydrogen/oxygen, although many of these pairs are of theoretical interest only:
Organic compounds/oxygen
Carbon monoxide/oxygen
Ammonia/oxygen
Hydrogen/chlorine
Metals/oxygen

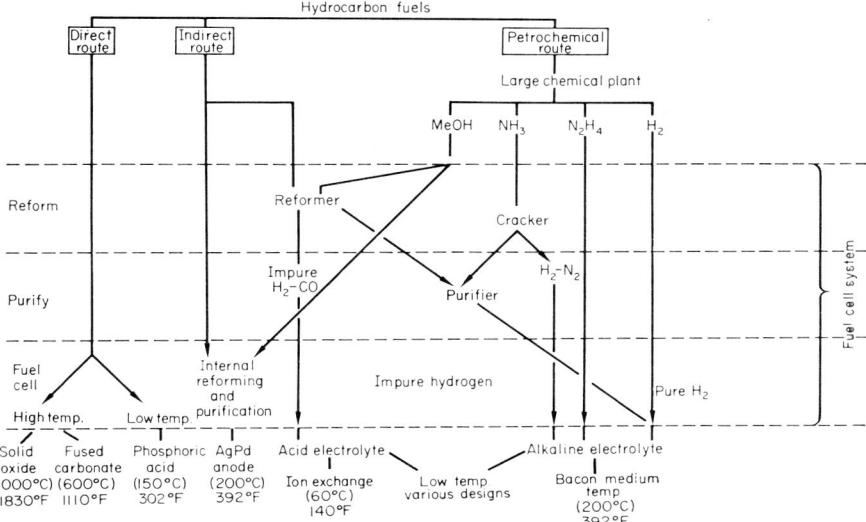

4.4 Basic fuel cell systems

Indirect fuel cells

These use a great variety of different fuels including coal gas, natural gas and oil. The first stage is the conversion of the feed into very pure hydrogen in a reformer. This hydrogen is then reacted with oxygen in a conventional hydrogen/oxygen fuel cell.

Regenerative fuel cells

The purpose of these is to store electricity. During off-peak periods chemicals are produced which can be converted back into electricity during periods of maximum power demand.

Hydrogen/oxygen fuel cells

Hydrogen gas can be produced either by a reforming process from hydrocarbon feedstocks, or by either electrolysis or indirect thermal cracking, using a nuclear reactor.

The reactions which take place in a hydrogen/oxygen fuel cell can be summarized as follows:

$H_2 \rightarrow 2H^+ + 2e$ (anode reaction)

$\frac{1}{2}O_2 + H_2O \rightarrow 2OH^- - 2e$ (cathode reaction)

$2H^+ + 2OH^- \rightarrow 2H_2O$ (water as waste product)

Hydrogen/oxygen fuel cells come in four main types:

Aqueous acid types

The electrolytes for such fuel cells are either diluted sulphuric or phosphoric acids, as well as such solid polymer electrolytes as fluorocarbons

4.5 Prototype 5 kW fuel cell. Courtesy Energy Conversion Ltd

with active sulphonic acid groups. Such a solid polymer electrolyte is marketed by Du Pont Inc. of the USA under the trade name of Nafion. Aqueous acid type cells are operated at between 60 and 190°C, depending upon type, and have overall electricity production efficiencies of a maximum of 30 per cent.

Originally the electrodes used consisted of noble metals, which contributed some 65 per cent of the total cost of manufacture of the cell. In addition, the noble metals were very subject to catalyst poisoning whenever the hydrogen used contained any carbon monoxide. Some of the newer systems use tungsten carbide for the anode electrode instead, which is both cheaper and not subject to carbon monoxide poisoning.

The Power Systems Division of the United Technologies Corporation is now building 26 MW fuel cell systems of this type. It is reckoned
that costs of phosphoric acid systems can be lowered to between $350
and $450 per kWe with an overall electricity power production cost of
between 4 and 5 cents per kWh.

Alkaline fuel cells
These use fairly concentrated solutions of aqueous potassium hydroxide or saturated carbonate/bicarbonate solutions as the electrolyte.
They can be operated as quite low temperatures (70°C to 200°C), have
efficiencies of conversion of the same order of magnitude as acid electrolyte cells, but cost up to 25 per cent more to buy and run. Alkaline
fuel cells do not need quite as pure hydrogen for feed, and electrodes
need not be made from precious metals: nickel has been found to be
quite suitable as a material.

Molten carbonate fuel cells
The electrolyte used in these cells is a mix of alkali carbonates and alu-
minates kept at a temperature of about 650°C. The anode is nickel,
while the cathode is normally nickel oxide coated with lithium compounds. High temperature operation (up to 750°C) improves the power
density of the cell but increases liability to corrosion.

Molten carbonate fuel cells cost roughly the same as alkaline fuel cells
per unit generating capacity, but their life expectancy is lower. So far
their longest stable good performance achieved between overhauls is
10 000 hours.

Fuel-electric power efficiencies with these types of cells can be as high
as 46 per cent. It is suggested that waste heat produced during operation of the cell is used for reforming purposes.

Stabilized zirconium fuel cells
In these high temperature fuel cells the electrolyte is stabilized zirconium oxide. This is a solid at the temperature of operation, which is
around 1000°C. A very high conversion efficiency of fuel to electricity
(approaching 60 per cent) is possible because activation polarization
losses are negligible. There are no liquids in this cell so that one avoids
all problems of pore flooding and maintenance of stable interfaces. On
the other hand, the electrolyte resistivity is very much higher than with
competing systems.

At the time of writing, the largest cell of this type made has a capacity
of about 100 W and a very short useful life. This cell may be quite
attractive in future because waste heat is produced at a high temperature, and it can be employed to gasify coal, the primary feed fuel. At the
present time it can be considered very much in the initial experimental
stage.

Reforming of fuel

When pure hydrogen is used as a fuel in alkaline fuel cells, one obtains 0.146 kWh of electricity per MJ of heating value. When impure hydrogen is used in an acid fuel cell this figure falls somewhat to 0.129 kWh. Hydrogen is, however, an expensive fuel to use and will be until we achieve the hydrogen economy (ie the replacement of present-day fuel resources in their entirety with hydrogen produced by nuclear reactors). Practical fuel cells therefore have to rely upon fossil fuels.

Coal, oil, natural gas and biomass are all sources of the primary gas to the fuel cell plant. Coal is nowadays converted into fuel gas by such means as the Lurgi complete gasifier, while fuel gases are produced both at the well-head of oil boreholes and as refinery by-products. One can use waste gases from sewage works and many other sources as the primary fuel feed. Catalyst converters are able to produce hydrogen gas from hydrocarbons and carbon monoxide.

At 870°C methane reacts with steam as follows:

$$CH_4 + H_2O \rightarrow CO + 3H_2 \qquad \Delta H = +226 \text{ kJ/mole}$$

The heat required for this endothermic reaction has to be provided from waste heat produced by the fuel cell.

On the other hand, the reaction

$$CO + H_2O \rightarrow CO_2 + H_2$$

is slightly exothermic and therefore self-sustaining. In a practical gas converter it is possible to produce approximately 3.8 m³ of hydrogen from each cubic metre of natural gas or refinery tail gas. In general, some 75–80 per cent of the hydrogen present in the hydrocarbon can be used in the fuel cell.

For most fuel cells, it is important to keep the carbon monoxide content down, as this affects the life of the fuel cell adversely. An efficient water gas converter is therefore an essential feature.

Operational considerations

The capacity of a given cell is increased considerably if one substitutes tonnage oxygen for air. On the other hand, this increases the operational cost appreciably.

Acid fuel cells have considerable tolerance for carbon dioxide but the various alkaline units need a feed gas reasonably free from CO_2. It is therefore necessary to install a CO_2 stripper consisting of a high pressure water scrubber combined with an alkaline absorber. This represents 28 per cent of the total capital costs of the fuel cell installation.

Fuel cells give their electric output in the form of very low voltage DC. The voltage is, of course, stepped up by connecting the cells in series but it has then to be converted to AC. Either line commutated or force commutated inverters are employed in the USA to provide 60 Hz AC power. Inverter sub-system losses vary between 4.5 and 6 per cent depending on the method used. In the operation of all fuel cells heat which should be utilized in some way is produced in the cell. Much of the heat is used for driving the converter and other auxiliaries, but this can usually only be done with high temperature equipment.

With the more common acid and base fuel cells, low grade heat is given off, which can best be used for space heating. So far, however, fuel cells suitable for domestic use have not been developed.

At present, fuel cells cannot compete with heat engine plants from a cost point of view. It is true that fuel cells can often use fuel more efficiently to generate current than heat engines can. The savings in prime fuel, however, do not counteract the very much higher capital cost of fuel cells and their very much lower expectation of life compared with the heat engines. It looks likely, therefore, that the use of fuel cells will be restricted for some time to circumstances in which they are competitive with heat engines because of their suitability for some specific requirement other than cost. Military use and space exploration uses are likely, then, to be the major outlets for fuel cell technology for some years to come.

SELECTION CHART: FUEL CELLS

Type of equipment	Advantages	Disadvantages	Appropriate applications
Hydrogen/oxygen acid cells	Operate at between 60 and 190°C 30% efficiency	Expensive materials of construction Needs pure hydrogen Subject to catalyst poisoning	Military, space, medical and other specialised uses up to 26 MW
Hydrogen/oxygen alkaline cells	Similar to acid cells Need less pure hydrogen than acid cells	Lower manufacturing costs but 25% higher operating costs than acid variety	Similar uses to acid type
Hydrogen/oxygen molten carbonate cells	Efficiency up to 46%	High operating temperature (750°C) Short life expectancy—less than 10 000 h	Suitable under circumstances where use can be made of waste heat in addition to DC electricity
Stabilized zirconium cell	Very high efficiency Absence of liquids	Operational temperature as high as 1000°C	Still only at prototype stage
Hydrocarbon/ oxygen cells	Cheaper fuel required than hydrogen	Fairly low efficiency Expensive materials of construction	Electrically powered vehicles such as fork-lift trucks, agricultural vehicles milk floats

References and further reading

1 R. Noyes, *Fuel cells for public utility and industrial power*, Noyes Data Corporation, Park Ridge, NJ, 1977.
2 J. P. Ackerman et alia, *Assessment study of devices of the generation of electricity for stored hydrogen*, Argonne National Laboratory, Illinois, 1975.
3 B. J. Crowe, *Fuel cells*, NASA Survey SP–5115, Washington, 1973.
4 J. O'M Bockris and S. Srinivasan, *Fuel cells; their electrochemistry*, McGraw-Hill, New York, 1969.
5 B. S. Baker, *Fuel cell technology*, Academic Press, New York, 1966.
6 J. Hammel, *Fuel cells*, Abacus Press, New York, 1966.
7 A. McDougall, *Fuel cells*, Macmillan, London, 1976.
8 W. Vielstich, *Modern processes for the electrochemical production of energy*, Interscience, New York, 1970.
9 D. Cameron, 'Fuel cell energy generators', *Chemical technology*, 9, October 1979, pp 933–937.
10 C. Sylwan, '50 MW Methane/air fuel cell', *Energy conversion*, 20, No. 1, 1980, pp 1–7.

SELECTED COMPANIES INVOLVED IN MANUFACTURE OF PRODUCTS

UK and Europe

Berec International Ltd, 1255 High Road, Whetstone, London N20 0EJ.
Chloride Automotive Batteries Ltd, Chequers Lane, Dagenham, Essex RM9 6PX.
Dornier System GmbH, Postfach 1360, 7990 Friedrichshafen 1, West Germany.
Hawker Siddeley Electric Ltd, 32 Duke Street, London SW1Y 6DG.
Union Carbide Ltd, High Street, Rickmansworth, Herts.
Varta Ltd, Hermitage Street, Crewkerne, TA18 8EY.

United States

A & M Engineered Composites Corporation, 3 Hayes Memorial Drive, Marboro, MASS.
ATL Inc., Spear Road, Industrial Park, Ramsey, NJ 07446.
Dayton Electric Mfg Co., 5959 Howard Street, Chicago, ILL 60648.
Electrochimica Corporation, 2485 Charleston Road, Mountainview, CAL.
Electrolyser Corporation, 120 The West Mall, Etabicoke, Toronto, Ontario, Canada.
Energy Research Corporation, 3 Great Pasture Road, Danbury CN.
Firestone Coated Fabrics Co., Firestone Drive, PO Box 887T, Magnolia, AR.
Fuel Cells Inc., 4278 Swinnea Road, Memphis, TN.
General Electric Company, Troup Highway, Tyler, TX 75711.
Goodyear Aerospace Corporation, PO Box 9278T, Akron, OH.
Tork Inc., 1 Grove Street, Mount Vernon, New York, NY 10550.

5 Heat pumps

Electricity is a high-grade fuel. Normally it is converted into heat by allowing it to flow through a nichrome resistance wire which is heated up by the current. The conversion of electric power into heat takes place according to the first law of thermodynamics, ie:
Electrical energy lost = heat energy gained.

One can, however, convert electricity into heat at a much better efficiency than that. One can use electricity to operate a compressor to liquefy a refrigerant such as freon, thereby yielding latent heat. This latent heat has now to be restored to allow the vapourization cycle to continue, so heat is abstracted from the surroundings at a very low thermal level. Using a heat pump one can get three times as much heat, or even more, for a given quantity of electricity than one can by simply allowing the electricity to flow through a nichrome wire. This is, briefly, the basis of the heat pump system. One does not get anything for nothing. The first law of thermodynamics is obeyed because the cold which is produced as the liquefied refrigerant evaporates constitutes a negative heat production. The sum still adds up:
Electrical energy = heat energy minus cold energy
The actual heat energy has therefore to exceed the heat energy produced by direct conversion of electricity to heat, as happens when one uses a resistance wire.
In a heat pump system we express the ratio:

$$\frac{\text{Heat energy produced}}{\text{Electrical or mechanical energy expended}}$$

as its coefficient of performance (COP).

The refrigeration effect of the heat pump can either be used as such to provide chilled air or cool down food, or if it is not feasible to employ such a system, it is simply neutralized by using some heat sink such as lukewarm water, the soil or external air in a second heat exchanger circuit.

The coefficient of performance is improved if the temperature of the cold side of the heat pump is raised. For this reason heat pumps have often been referred to as devices able to convert low grade thermal energy into useful heat.

5.1 The principle of the heat pump. Courtesy Happel GmbH
The heat pump uses the refrigeration cycle to extract heat from air or water
by using a fluid refrigerant which is at a lower temperature. A compressor is
used to pump the refrigerant round a circuit, as shown in the diagram. Heat
is removed from air or water by the refrigerant as it flows through the
evaporator coil. The absorption of heat by the refrigerant changes its state
from a cold liquid/vapour mixture to a cool gas which is then drawn into a
compressor in which its pressure and temperature are increased. This
increase in temperature allows heat to be transferred from a condenser coil
direct to air or water. The cooling effect of the condenser converts the gas to
a hot liquid still under high pressure and by means of a thermal expansion
valve into a low pressure mixture of cold liquid and vapour. The cycle is
then ready to be repeated.

GENERAL PRINCIPLES OF HEAT PUMPS

The heart of the heat pump is the compressor. This can be driven
either by an electric motor, by wind or water power, or by a petrol, gas
or diesel engine. A refrigerant is in the form of a gas when it flows into
the compressor. The work done in compression increases its density
and it also heats up appreciably. The hot compressed gas then flows
into the condenser, which is an efficient heat exchanger, where the
excess heat is removed. As the gas condenses to form a liquid, its latent
heat is used to heat up either water or air via the heat exchanger.

The nature of the refrigerant is of some importance. It is essential to
avoid fluids where the temperature at which heat is taken off is above
the fluid's critical point. The critical point is defined as the temperature
above which a gas cannot be liquefied by pressure alone. For example, a
normal heat pump could not use either oxygen or nitrogen as a
refrigerant medium because the temperature of the hot section is very
much higher than the critical point of either of these two gases.

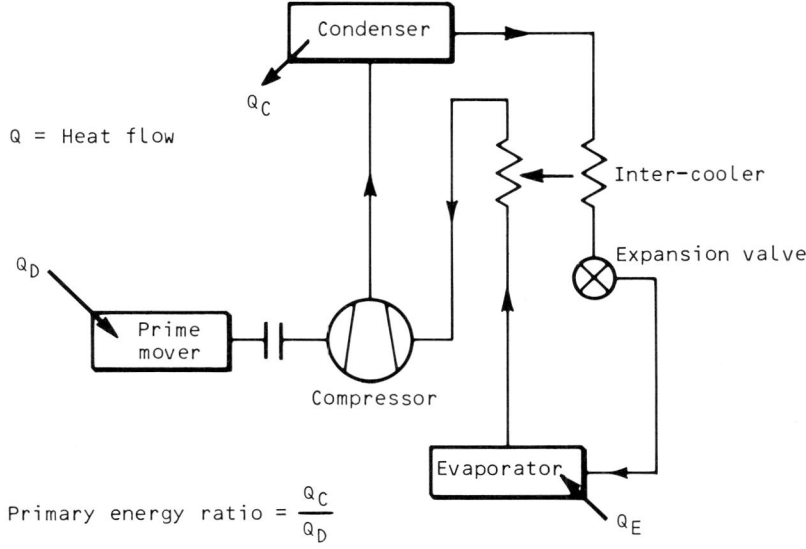

Q = Heat flow

Primary energy ratio $= \dfrac{Q_C}{Q_D}$

A

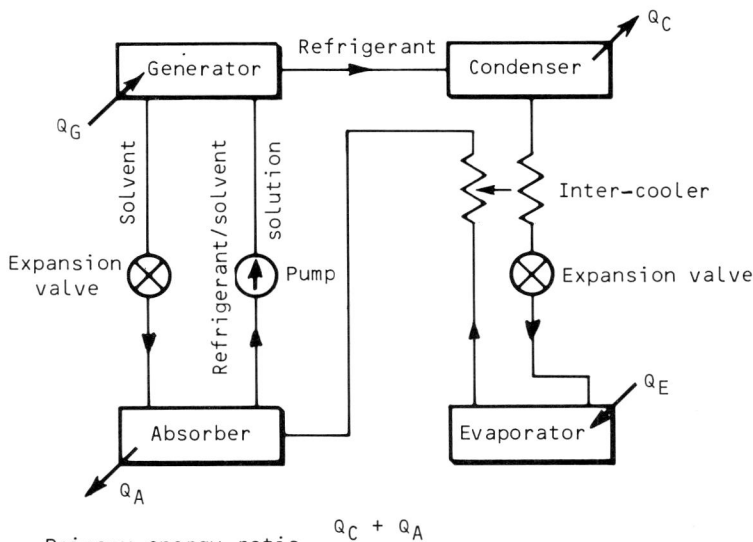

Primary energy ratio $= \dfrac{Q_C + Q_A}{Q_G}$

B

5.2 Heat pumps: A: Rankine cycle   B: Absorption cycle   Courtesy V. A. Eustace and A. Wright, and International Research and Development Ltd

For optimum performance, it is desirable to take off as much heat as possible in the condenser/heat exchanger. The liquid which has now had its excess heat removed next flows to the expansion section, where it evaporates. As this evaporation takes place, a refrigeration effect is produced due to the need for a liquid to absorb latent heat as it vapourizes.

The theory of the heat pump in detail

When a gas is compressed by a piston it heats up. This heat can be taken off by a heat exchanger so that the compressed gas returns to its original temperature. If the compressed gas is now allowed to expand, its temperature falls and further heat has to be supplied to it from an extraneous source, which in heat pump language is called the heat sink. In actual fact the operation of a heat pump using a simple gas would be very uneconomical. One uses a refrigerant instead, which is a gas that condenses to a liquid when subjected to pressure and cooled, subsequently reverting to the gaseous state when the pressure is released.

When a gas changes into a liquid, a considerable amount of heat, called latent heat, is evolved. For example, the latent heat of refrigerant 12 (CCl_2F_2) is 165.22 kJ/kg. The same amount of heat is absorbed from the surroundings when the liquid changes back again into vapour. Thermodynamically speaking, a heat pump is the reverse of a heat engine, in which mechanical energy is produced by utilizing the difference between two temperatures. In the heat pump, however, mechanical energy is fed in to produce two different temperature levels.

If the system is to be operated simply as a heat pump and not as a refrigerator, low grade heat is supplied to the fluid from the surroundings via an efficient heat exchanger. It is necessary to bring up the temperature of the gas to the inlet conditions, which exist at the entrance to the compressor. The heat supply at the cold end of the heat pump can be at quite a low temperature, ie one can use atmospheric air blowers, running river water, ground water, sea or lake water, provided its temperature is higher than that of the evaporated fluid.

Heat pumps which use very low grade heat on the cold side do not, however, operate with a particularly good coefficient of performance (COP). It is better to use instead somewhat warmer waste heat such as condenser water from power stations, or lukewarm water from industrial processes, as the COP is raised appreciably under such conditions. A popular source of low-grade energy which can be upgraded by heat pumps is solar heat. It is used to warm up water reservoirs which then supply the low grade heat needed by heat pumps. Heat pumps can also be employed to upgrade geo-thermal energy.

Naturally, it is always best when one can use the refrigeration effect for some purposes as well. Under such circumstances the low grade heat needed is supplied by the substances which have to be refrigerated: meats and other food which have to be preserved by cooling, or water which has to be chilled for air conditioning, or to enable it to be turned into ice. Obviously it is highly advantageous if heat pumps can

HEATING MODE - HEAT PUMP

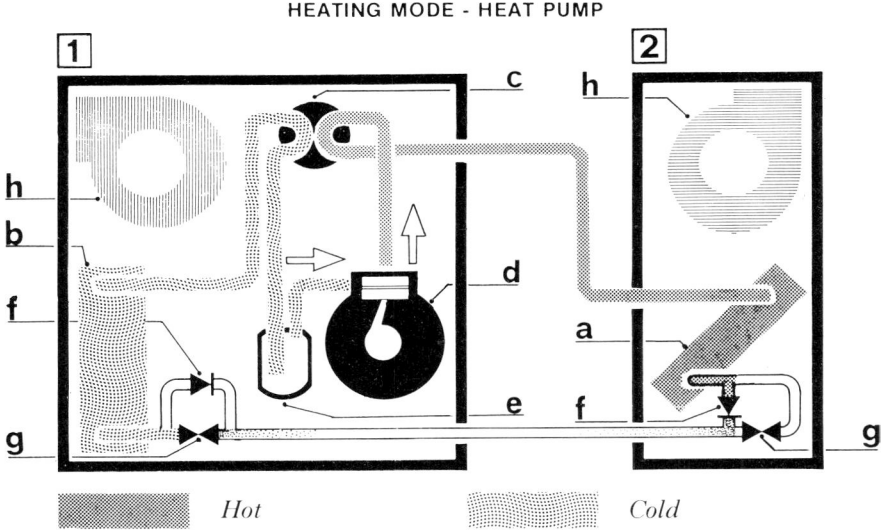

Hot   *Cold*

COOLING MODE - AIR CONDITIONER

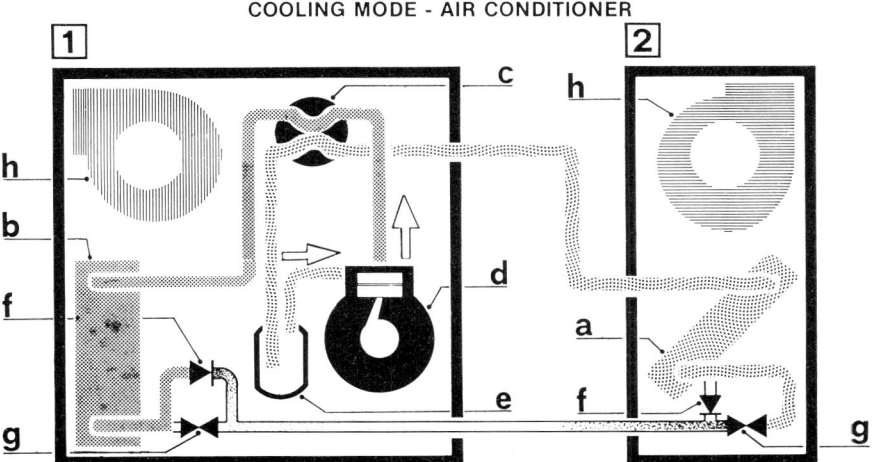

5.3 Operation of heat pump in heating and cooling modes. Courtesy F. H. Biddle Ltd

1 Condenser	*c Cycle-inversion valve*	*g Thermostatic valve*
2 Air handler	*d Compressor*	*h Fan*
a Inside exchanger	*e Liquid separator*	
b Outside exchanger	*f One-way valve*	

be used in this way, because the same equipment can perform two separate duties for the same capital and running costs.

In recent years heat pumps have come down considerably in price, because of the introduction of mass production methods into the manufacture of compressors and auxiliaries.

As the cost of primary fuels increases, heat pumps which can upgrade low temperature heat into usable heat have shown themselves increasingly viable. A relatively low expenditure of prime fuel can provide adequate space heating, with valuable refrigeration capacity.

Appropriate conditions for using heat pumps

Heat pumps have been found particularly suitable under the following circumstances:

When cheap electricity is available

Much less electricity is normally consumed at night than during the day. While it is possible to throttle down fossil fuelled plants fairly simply, this cannot easily be done with either hydroelectric equipment or with nuclear power stations. Both can only be operated most economically if run at full load. As nuclear electricity costs less than electricity obtained by burning fossil fuels, nuclear power stations tend to carry the fairly constant base load of electrical capacity. Fossil fuelled plant provide the variable load. Even now, however, it pays to sell electricity at a considerable discount for night-time use, amounting in most cases to a figure of around 50 per cent. As the percentage of electricity produced by nuclear energy increases, it will become ever more necessary to find a commercial outlet for electricity produced at night. It is therefore likely that off-peak electricity will be sold even more cheaply than today. The heat pump enables this electricity to be converted into heat at an efficiency of 200—300 per cent and heat, unlike electricity, can be stored against peak-time requirements.

When natural gas is available

Natural gas is a cheap fuel. It can be used for heating either by direct combustion or by use in an internal combustion engine coupled to a compressor and heat pump. In such a system, the overall heat given out exceeds the calorific value of the gas fed in. One uses for space heating the exhaust and jacket heat of the gas engine, as well as the heat given out by the compressor. In addition, the system yields refrigeration energy. It is possible to justify gas driven heat engine/heat pump systems for the purpose of supplying only heat either in the domestic or in the industrial context.

Where refrigeration energy is used

A heat pump is easiest to justify when refrigeration energy is needed for such purposes as space air conditioning, freezing of food or other industrial processes. For example, sports complexes including an indoor ice skating rink can be heated up most economically from the compressor heat of the refrigeration machinery. Heat pumps are commonly used in shopping precincts where large quantities of chilled water are required, together with space heating for offices and other premises.

Where very cheap low grade heat is available
Heat pumps can be used to up-grade waste heat which is at too low a temperature to be useful itself, but which can be raised to a suitable level by expending a limited quantity of electricity.

Such very cheap low-grade energy can be in the form of solar heat, waste heat from certain industrial processes, condenser water from power stations, or geothermal heat. Since the up-grading process has to be carried out by very costly electricity, it is necessary to make a very accurate economic assessment first. In the absence of a reasonable refrigeration load it is frequently difficult to make such a system pay if one has to use on-peak electricity.

Using unconventional fuels
Research is at present underway in many countries on the use of fuels such as low-grade coal, refuse, etc. for running stirling engines to pro-vide electric power. Systems of this type could be used with heat pumps to obtain space heating energy and refrigeration energy.

For space heating/cooling in countries with appreciable differences between external winter and summer temperatures
In most technically developed countries one requires heating during the winter months and air conditioning in summer. Traditionally one employs steam heating during the cold months of the year, using a sim-ple electrically driven air chilling aggregate during the summer. Obviously such a system has considerable disadvantage as against a heat pump, which can work in both a heating and a cooling mode. Simple valve operation can be adopted to enable the same plant to supply both heat and cold, simply by interchanging the roles of two heat exchangers to function as condensers and evaporators, respectively. Modern equip-ment of this type can be operated either manually or automatically using a relay activated by external ambient air conditions and internal desired temperature.

The Carnot equations
As is well known, any heat engine such as a motor car engine or steam turbine has a thermal efficiency which is limited by Carnot's equation:

$$\text{Efficiency} < \frac{T_2 - T_1}{T_2}$$

where T_1 is the lowest temperature in the system, ie the exhaust
temperature in kelvins
T_2 is the highest temperature in the system, such as the feed-in

saturated steam temperature in a turbine, or the temperature of the air/petrol mixture at the explosion temperature, again in kelvins.

In practice, thermal efficiencies of even the most efficient steam turbines are unlikely to exceed about 36 per cent and there are few designs of internal combustion plant with a thermal efficiency above 40 per cent. A heat pump is the reverse of a heat engine and therefore one expresses the COP as

$$COP < \frac{T_2}{T_2 - T_1}$$

which is called the reciprocal Carnot equation.

Once more T_2 and T_1 stand for the highest and the lowest temperature in the system, in kelvins.

Unfortunately, practical COP values bear little resemblence to theoretical Carnot COP values; for the same reasons that practical efficiencies of prime movers are nowhere near the theoretical efficiencies calculated from the Carnot equation. The reasons for this are fourfold:

1 *Inefficiency of heat exchangers*

Heat pumps require two heat exchangers, one at the hot end and one at the cold end. In both cases a temperature drop must exist for heat to pass from one fluid to the other. This means, in practice, that the temperature of the primary fluid of the heat pump which supplies the heat has to be some 15°C higher than the supply temperature. Equally, the cold end which absorbs thermal energy from the so-called heat sink has to have a temperature some 15°C lower than the heat sink. Obviously, the better the heat exchanger, the lower the temperature difference across it is going to be. Equally, if the thermal conductivity across the heat exchanger deteriorates, as can happen when surfaces are subject to scaling or are otherwise fouled, then the temperature difference may rise well above the figure of 15°C given, thereby reducing the COP of the heat pump even further.

2 *Isentropic efficiency*

An ideal compressor would be one in which there is no entropy change whatsoever between the feed refrigerant and the compressed product pumped out at the other end. This means that an ideal compressor would have to be a perfectly reversible engine. This cannot be done in practice as there is always a positive entropy change in the thermodynamic process, so that the compression process suffers irreversibility to some extent. This is expressed by the term isentropic efficiency, which can be calculated from the following equation:

$$\text{Isentropic efficiency} = \frac{H_{dis} - H_{feed}}{H_{th} - H_{feed}}$$

where H_{dis} is the discharge enthalpy of the refrigerant in kJ/kg
  H_{th} is the theoretical discharge enthalpy of the refrigerant in kJ/kg assuming that the compression process takes place with zero entropy change
  H_{feed} is the enthalpy of the refrigerant fed into the compressor in kJ/kg.

An average isentropic efficiency for a heat pump compressor is about 0.75.

3 Sensible heat loss of refrigerant

A further source of inefficiency in a heat pump system is caused by the fact that the fluid which exits from the condenser/heat exchanger is at the temperature at which it has given off heat to the secondary fluid of the heat supply system. Although all the latent heat of condensation has been used, the fluid still carries an appreciable amount of sensible heat when it enters the expansion unit. When single stage compression is used, this lowers the overall efficiency of the heat pump quite appreciably.

In modern multi-stage equipment the sensible heat is used in flash chambers to permit partial evaporation of refrigerant at higher temperatures, followed by recompression in the higher stages of the multiple compressor system. This partial re-use of sensible heat is the reason why multi-stage heat pumps have a better COP than single-stage units.

Sensible heat losses vary with the temperature difference between the two heat exchangers. For example, if the temperature difference is 50°C the COP can be affected by a factor of about 0.7 by this agency. If the temperature difference is only 25°C the factor to be applied is 0.85. As a general empirical rule, the factor by which the COP is reduced as a result of sensible heat losses is:

$1 - 0.006 \, (T_c - T_e)$

where T_c is the condensation temperature in kelvins
and T_e is the evaporation temperature in kelvins.
The use of multi-stage heat pumps with several compressors increases this factor to:

$1 - 0.003 \, (T_c - T_e)$ for two-stage sets and to
$1 - 0.002 \, (T_c - T_e)$ for three-stage sets.

4 Machine efficiency

Electric current is used to drive the compressor, and there is inevitably some wastage of electric power in the windings and commutator of the motor. Further irreversible heat losses occur in the bearings of the motor and compressor, while auxiliaries such as pumps and valves also induce certain machine losses. Really well-designed, brand new equipment may have a machine efficiency as high as 0.92, but this is rather an optimistic estimate for long-term operation. For plant which has been installed for some years the machine efficiency is unlikely to be much higher than about 85 per cent or 0.85.

Calculation of theoretical and practical COP for a single compressor heat pump system

Let us consider a heat pump system which supplies hot water at 80°C at the condenser end, using a sink temperature of 35°C for its evaporator. The heat exchangers on both the hot and cold side need a temperature difference of 15°C, and the isentropic efficiency of the compressor equals 75 per cent. The machine factor of the equipment is 0.90. Calculation of the theoretical COP is easy:

It is simply $\dfrac{(80 + 273)}{(80 - 35)} = 7.845$

To evaluate the practical COP we must take several factors into consideration:

Inefficiency of heat exchanger devices
To operate the heat exchangers properly the condensation temperature must be 80 + 15 = 95°C, while the evaporation temperature equals 35 − 15 = 20°C.
We can therefore calculate the reverse Carnot COP to be

$\dfrac{(95 + 273)}{(95 - 20)} = 4.907$ and not 7.845

Isentropic efficiency
A factor of 0.75 is introduced to allow for the fact that the compressor operates with an increase in entropy.

Sensible heat losses
The difference between the outflow from the hot heat exchanger and the inlet to the cold heat exchanger is approximately equal to the difference in supply temperatures, namely 80—35 = 45°C, assuming that one uses counter-current heat exchange. If one uses co-current heat exchange, the temperature difference is a good deal higher.
Substituting into the equation for the sensible heat factor for single stage pumps we get:
Sensible heat loss factor = 1 − (0.006 × 45) = 0.73.

Machine factor
The machine factor is 0.9.
From these data it is now possible to calculate the practical COP for the equipment.

It is equal to: $4.907 \times 0.75 \times 0.73 \times 0.90 = 2.42$.
If the plant had been run with twin compressors in series and a flash chamber in between, the sensible heat loss factor would have been:
$1 - (0.003 \times 45) = 0.865$
In consequence the practical COP of the unit would have been:
$4.907 \times 0.75 \times 0.865 \times 0.9 = 2.86$

Many heat pump manufacturers tend to quote a given COP for their equipment without specifying the temperature limits between which the equipment is intended to be used. This is, of course, totally valueless, as the difference between input and output temperatures is very vital in calculation of the COP.

Table 5.1 gives the practical COPs for a supply temperature of 80°C and a temperature drop in both heat exchangers of 15°C, assuming as before an isentropic efficiency of 0.75 and a machine efficiency of 0.9. Practical COPs are given for single stage, double stage and triple stage compressor systems.

TABLE 5.1 PRACTICAL COP FIGURES
Figures are for single, double and triple stage heat pump systems using varying heat sink temperatures. The supply temperature is 80°C

Heat sink temperature	Practical COP		
	Single stage	Double stage	Triple stage
70	5.84	6.02	6.09
65	5.02	5.27	5.35
60	4.37	4.67	4.77
55	3.84	4.18	4.29
50	3.39	3.77	3.89
45	3.02	3.42	3.55
40	2.70	3.12	3.26
35	2.42	2.86	3.01
30	2.17	2.64	2.79
25	1.96	2.44	2.60
20	1.77	2.26	2.43
15	1.59	2.10	2.27
10	1.44	1.96	2.14
5	1.30	1.83	2.01

Equally, the COP drops if one has to supply the heat at a higher temperature. For example, when the sink temperature is 35°C the practical COP equals 2.42 if a single compressor is used and one seeks to supply the heat at 80°C. If one needs to supply heat at 100°C instead, the COP drops to 1.68.

The primary energy ratio PER

The COP assumes that the compressor is driven either by electricity or some other form of mechanical energy, such as shaft energy from a water wheel or wind powered appliance.

Many heat pumps use a basic fuel such as gas or oil. In the future hydrogen, produced either by electrolysis or indirectly by thermal cracking of water using the heat developed by a nuclear reactor, looks like being a most important fuel for driving heat pumps. The PER is defined as:

$$PER = \frac{\text{Useful heat delivered by the heat pump}}{\text{Heat equivalent of primary energy used}}$$

If one employs the primary fuel in a heat engine with an efficiency of EFF, then the primary energy ratio PER is given as the product of EFF and COP, together with heat recoverable from the operation of the heat engine.

Example: If one employs a gas engine with an efficiency of 41 per cent (EEF = 0.41) and couples it to a heat pump with a COP of 3.2, then the net primary energy ratio $PER_{net} = 0.41 \times 3.2 = 1.312$.

It should be possible to recover a further 35 per cent of the input primary fuel energy to the gas engine in the form of exhaust and jacket heat. The gross PER is then $1.312 + 0.35 = 1.662$.

If the gas is used for firing a boiler directly, it can probably be used with an efficiency of around 80 per cent. It can be seen that a gas- or oil-fired heat pump system has more than twice the efficiency for heating as compared with using the fuel directly in a boiler. In addition one can, of course, obtain the benefit of refrigeration capacity when one employs a heat pump. At present prime fuels are still sufficiently cheap to make the substitution of a heat pump for a boiler installation a matter of doubtful viability if one merely seeks to get one's heating from either gas or oil in the most economic way. Gas or oil-driven heat pumps are exceedingly viable for use when both heat and cold have to be supplied.

Multi-stage heat pump system

This is analogous to the re-heat cycle which is employed today with all large steam turbines. One uses two or more compressors in series. The compressed gas passes into the condenser where heat is given off as it liquefies. The liquid flows through the first expansion valve, so that partial evaporation occurs, with some of the vapour flashing off. This flashed-off vapour is recompressed in the high pressure stage of the compressor only. In multi-stage heat pump systems this process is repeated down the line.

The COP of such a system is improved because one can make use of the sensible heat which remains after the condensation heat has been given off. Multiple stage heat pumps are only used for very large installations. In smaller units the additional cost of equipment cannot be justified by the small improvement in COP obtained. The specific heat of refrigerants is comparatively slight in comparison to the latent heat. For example, refrigerant 12 (CCl_2F_2) has a latent heat of 165.22 kJ/kg, while its specific heat is only 0.97 kJ/kg K.

The COP loss is conditioned by the difference in the temperature at which the liquid refrigerant exits from the condensing coil and that at which it emerges from the heat sink coil. The greater this temperature difference the better the improvement of the overall COP when using multiple compressor systems.

Refrigerants

These are liquids which can vapourize with an appreciable absorption of latent heat, which is given out again when the vapour is reconverted into a liquid. Many different refrigerants are used in the heat pump industry, the choice depending upon the temperature range being covered, the compression ratio used, and several other factors.

Although one uses fluids such as water, ammonia and sulphur dioxide in some heat pumps, the most usual refrigerants are a series of hydrocarbons where one or more of the hydrogen atoms have been replaced by halogens such as fluorine, chlorine and/or bromine. These refrigerants bear such trade names as freon, arcton, carrene, etc.

In addition, refrigerants are classified by a numbering system universal in the industry. The rules governing this system are given below:

a The first digit is the number of carbon atoms in the molecule minus 1.

b The second digit gives the number of hydrogen atoms in the molecule plus 1.

c The third digit gives the number of fluorine atoms in the molecule.

d Mixtures of refrigerants are designated by special numbers starting with 500, eg $CCl_2F_2/CH_3CHF_2 = 500$.

e Oxygen and nitrogen compounds start with 600, eg diethyl ether = 610.

f Inorganic refrigerants such as ammonia or sulphur dioxide start with 700, eg $NH_3 = 717$.

g Unsaturated organic compounds are designated by an additional 1 in front of the first digit, eg trichloroethylene $CHCl:CCl_2 = 1120$.

To show how the general rule of the numbering of refrigerants works, let us take the refrigerant chloropentafluoroethane $CClF_2CF_3$ as an example:

There are two carbon atoms so that the first digit is:

$2 - 1 = 1$

There are no hydrogen atoms, so the second digit equals $0 + 1 = 1$. There are five fluorine atoms, so the third digit equals 5. The reference number for this refrigerant is therefore 115.

TABLE 5.2 PROPERTIES OF SELECTED REFRIGERANTS USED IN HEAT PUMPS

Reference number	Formula	Molecular weight	Boiling point °C	Freezing point °C	Critical temperature °C	Latent heat kJ/kg
11	CCl_3F	137.4	23.8	−111.0	198.0	180.42
12	CCl_2F_2	120.9	−29.8	−158	112	165.2
13	$CClF_3$	104.5	−81.4	−181	28.9	148.5
14	$CHCl_2F$	88.0	−128	−184	−45.7	136.0
21	$CHCl_2F$	102.9	8.9	−135	178.5	242.2
22	$CHCl_2F_2$	80.5	−40.8	−160	96	233.6
23	CHF_3	70.0	−82	−155.2	25.9	239.6
112	$C_2Cl_4F_2$	203.8	92.8	26.0	278.0	154.9
113	$C_2Cl_3F_3$	137.4	47.6	−35.0	214.1	146.8
114	$C_2Cl_2F_4$	170.9	3.8	−94.0	145.7	136.1
115	C_2ClF_5	154.5	−38.7	−106	80.0	126.0
116	C_2F_6	138.0	−78.2	−100.6	19.7	117.1
717	NH_3	17.0	−33.3	−77.8	132.8	1370.7
718	H_2O	18.0	100.0	0.0	374.2	2257.0
764	SO_2	64	−10.1	−72.7	157.1	388.8

The use of absorption heat pumps

Absorption refrigeration plant is employed to avoid installing a mechanical compressor. The COP is much lower than with compressor-driven heat pumps, but it is possible to use high grade heat in order to produce medium grade heat together with some refrigeration.

The basic theoretical Carnot COP of an absorption refrigeeration system can be quoted as:

$$COP_{th} = \frac{T_c T_g - T_e T_a}{T_c (T_c - T_e)}$$

where T_c is the temperature inside the condenser/heat exchanger in kelvins.

T_e is the temperature inside the evaporator/heat exchanger in kelvins.

T_a is the temperature in the absorber in kelvins.

T_g is the temperature in the generator in kelvins.

The way an absorption refrigeration system works is as follows. A concentrated solution of refrigerant in the carrier liquid is pumped to the heating section. Heat causes evaporation of the refrigerant and its compression, while the depleted liquid passes to a heat exchanger to give off its sensible heat, and then returns to the absorber vessel. The

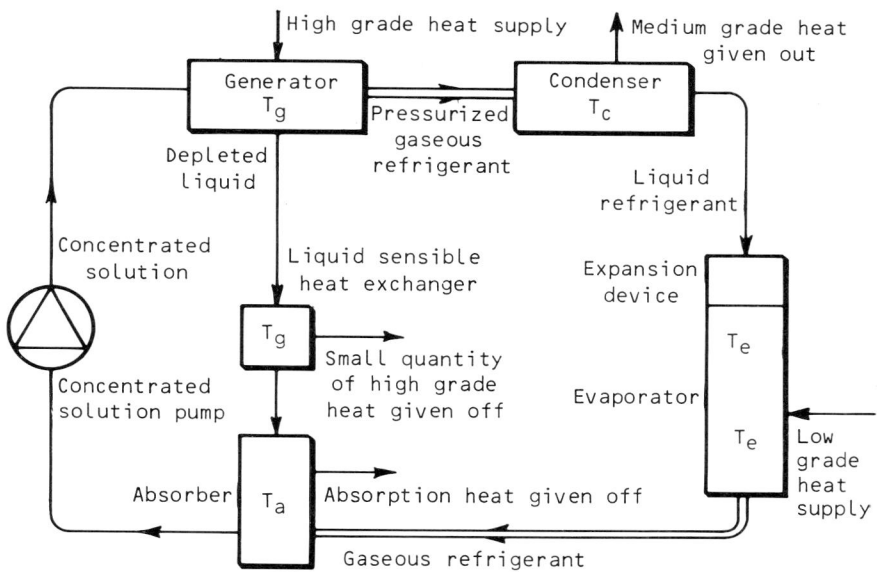

5.4 Absorption heat pump

compressed gas is piped to the condenser, where it gives off condensation and sensible heat. The liquefied refrigerant then passes to the expansion device where, as with normal compression heat pumps, it turns into a gas, with a consequent appreciable drop in temperature. To restore the temperature of the cooled gas to ambient levels, one can either use the cold generated for refrigeration, or where this is not feasible employ some low grade source of heat. The liquid which has been cooled down to an ambient temperature then enters the absorber and meets the vapour, which is also at ambient, to produce once more the concentrated solution of refrigerant which is pumped to the generator for heating and separation of the refrigerant. **Figure 5.4** shows the circuit employed. The refrigerant solution is normally pumped round the circuit by a very small electrically driven pump. The electric power needed for this is negligible in comparison to the overall quantities of heat involved in the system. It is also possible to use a thermosiphon system employing a neutral gas such as nitrogen, but this is usually only used in systems where for some reason or other it is impracticable to have the unit connected to mains or battery power. Two common pairs of refrigerants/liquids absorbants are used:

Ammonia as the refrigerant with water as the absorbent.

Water as the refrigerant with lithium bromide as the absorbent.

Coefficients of performance with absorption heat pumps are on average, about 40 per cent of those appertaining to equivalent compression heat pumps. The reason for this is that there are more heat exchangers with such a system, with consequent superheat losses. Uses of absorption heat pumps tend to be very limited.

Vapour recompression

Vapour recompression is a technique used to recover energy from commercial evaporation systems such as sugar evaporators, sea water desalination plant and the like. Significant energy savings are possible, as the mechanical energy supplied only delivers the difference in latent heat between the low grade output steam and the high grade input steam. This is small in comparison with the latent heat of the steam at the lower temperature, which would otherwise be wasted.

For example, let us assume that water vapour is given off at 50°C. At this temperature its enthalpy equals 2.592 MJ/kg.

The enthalpy of steam at 100°C is equal to 2.675 MJ/kg. It can therefore be seen that the expenditure of a mere (2.675 − 2.592) = 0.083 MJ = 83 kJ of compression energy can theoretically upgrade 1 kg of useless water vapour at 50°C to potentially useful steam at 100°C.

In practice, obviously, vapour recompression does not yield quite such good results, but experience has shown that for multiple industrial evaporator systems, vapour recompression can reduce primary consumption by as much as 80 per cent.

HIGH TEMPERATURE ABSORPTION CYCLES FOR POWER GENERATION TOPPING

Absorption cycle machines have no compressor and it has been suggested that this principle may be adopted to improve the thermodynamic efficiency of electric power generation. It involves the upgrading of the condenser water, which is at a temperature too low for practical purposes, to a temperature at which the steam from it can be fed into the low pressure section of the turbine. One would use some high pressure steam to carry out the upgrading process.

Assuming that large absorption heat pumps can be built with the same COP as the small units currently used for domestic or small industrial purposes, it is claimed that the efficiency of power generation of steam turbines can be increased from about 35 per cent to about 50 per cent.

As in a fossil-fuelled steam power station fuel costs amount to about 67 per cent of the annual expenditure, while capital costs only account for around 25 per cent, it can be calculated that a system able to save say 33 per cent of the fuel needed could cost roughly the same as the power station itself and still be a viable proposition.

Although some experimental work has been carried out in this field in West Germany, no plant of this type has yet been built.

DOMESTIC HEAT PUMPS

Normal daytime electricity tends to cost up to three times as much as natural gas per unit heat output, and thus the heat pump has to have a COP of at least 3 to compete in running costs with a simple gas-fired central heating system. As a heat pump is more expensive to install than a direct gas burner, it is not viable under such circumstances.

Domestic heat pumps can however be justified on the following grounds:

a If the purpose of the heat pump is dual, ie in winter it provides space heating, using the ground or external air as a heat sink, while in summer it provides room cooling. It is best to combine the heat pump with an energy storage device to obviate the use of costly daytime electricity.

b While in urban areas gas is piped to most dwellings, this is not the case in rural areas. Electricity is usually far cheaper to distribute than gas. Coal and oil have some disadvantages for space heating. The former is dirty and laborious to handle, while suitable heating oils are costly. In such a case heat pumps are direct competitors with normal electric heating: as a bonus one obtains a high heat output caused by the COP, as well as refrigeration capacity. Heat pumps are therefore of particular value to the farming community.

c Solar heaters have the disadvantage that heat is being collected part of

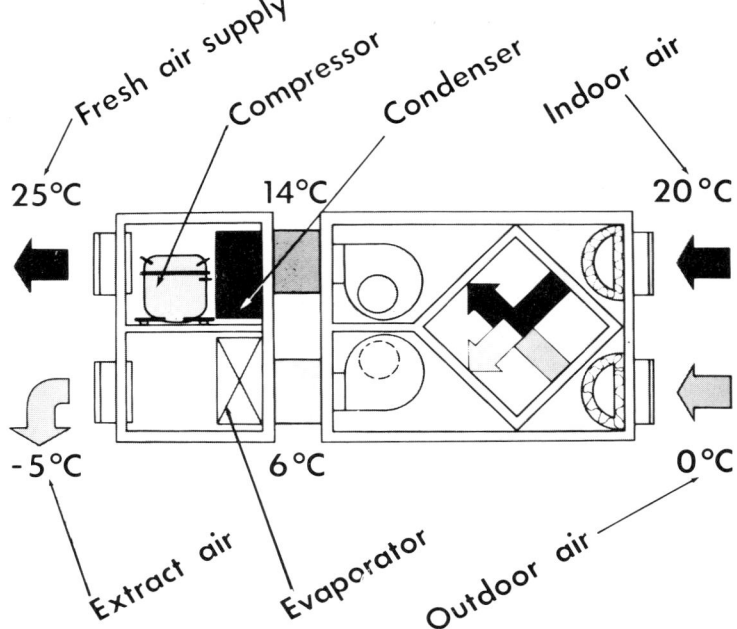

Fresh air supply

Compressor

Condenser

Indoor air

25°C   14°C   20 °C

−5°C   6 °C   0°C

Extract air

Evaporator

Outdoor air

5.5 Layout of domestic heat pump system. Courtesy F. H. Biddle Ltd

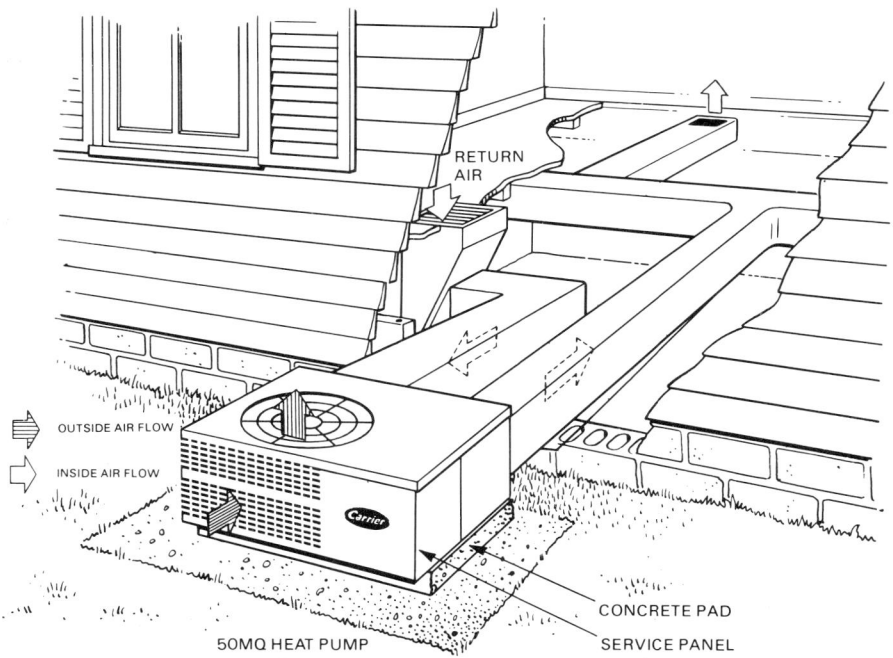

RETURN
AIR

OUTSIDE AIR FLOW

INSIDE AIR FLOW

Carrier

50MQ HEAT PUMP

CONCRETE PAD

SERVICE PANEL

5.6 Carlyle domestic heat pump installation

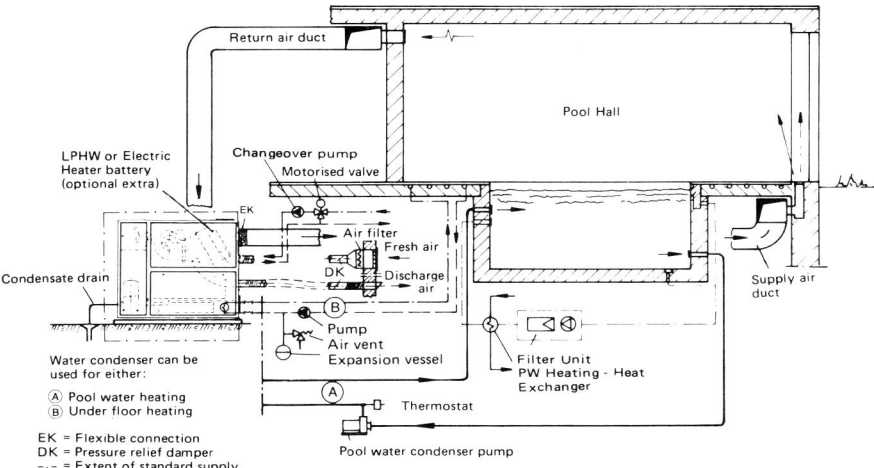

5.7 Fritherm heat pump system used for swimming pool heating

the year and part of the day only, and is necessarily low grade. Provided solar heating is combined with a suitable energy store, the low grade energy held in it can be used to improve the COP of a heat pump very markedly. Under such circumstances a good return can be obtained even when an appreciable fraction of the electricity used is charged at daytime rates.

d A system of domestic heating where heat pumps are particularly viable is one where heat has to be recovered from the exhaust air. One of the best ways of doing so is to use the exhaust heat exchanger to supply low grade heat to the evaporator section.

e Heat pumps are used in all types of experimental schemes. Prime mover energy comes from sources such as waterwheels, windmills and the like. Such schemes tend to be idealistic self-sufficiency exercises rather than practical applications. There are, however, occasions when they are economically viable.

f Heat pumps can be made viable when night-time electricity is being sold at a considerable discount. Under such circumstances water can be heated at night using cheap electricity and a favourable COP. This is stored in well insulated tanks. The hot water can be used for space heating and consumption purposes on the following day.

Although at the present moment it is difficult to justify installing heat pumps in dwellings except under the conditions mentioned above, the long-term future for domestic heat pumps is quite excellent. Full air conditioning, which necessitates air chilling for part of the year at least, is already the standard in the United States, parts of Canada, parts of Europe and Japan. Heat pumps are infinitely more economical than directly fired room heaters for winter and separately installed window air conditioning units for summer.

Throughout the world an increasing proportion of the general power supply is generated by nuclear power. The characteristics of nuclear power are that capital costs dominate while fuel costs are very low. In addition, it is difficult to throttle down nuclear power stations during periods of low power demand.

It is virtually certain that the difference between on-peak daytime electricity and off-peak night electricity prices will increase markedly in the future, to encourage more use of power at off-peak periods.

Design of domestic heat pumps

Many commercial heat pump systems are specially designed to supply dwellings. The Riello system marketed in the UK by Biddle Ltd is typical. For a small house with a heating capacity of about 10.6 kW one would use a compressor driven by a 3.5 kW motor. Hot water is circulated to the radiators of the house through narrow bore piping at a rate of 1300 litres per hour and a temperature of roughly 50°C. To keep the evaporator section at a reasonably steady temperature, a heat exchanger is placed deep down in the ground where the temperature is at about 8°C all the year round. Such a heat pump can be operated in either heating or cooling modes.

In the *heating mode* the water from the evaporator is pumped through the heat sink (the heat exchanger installed deep underground). Although the temperature of the heat sink is only 8°C a COP of 3 can be maintained for the outflow of water from the condenser, which is used for the supply of heat to the dwelling.

In the *water chilling mode* the heat pump is now operated so that the water from the condenser is pumped to the heat sink and gives off its heat to the soil. The water from the evaporator is pumped to a heat exchanger to supply chilled water or chilled air for indirect or direct space cooling purposes.

Such dual purpose space heating/space cooling heat pumps are often switched entirely automatically by room thermostats.

Domestic heat pumps can use other sources of waste heat besides the ground heat sink described. If a ventilation system is fitted, it is convenient to extract heat from the stale exhaust air, using either a run-around heat exchanger, a heat wheel or a heat pipe system.

Domestic swimming pools make excellent heat sinks. In winter, when they are not used, they serve as suppliers of low-grade thermal energy for the evaporator of the heat pump. In summer, the temperature can be raised somewhat by passing water from the condensing side of the heat pump through the swimming pool heat exchanger using the cold produced at the evaporator end for room chilling. One may also employ simple air/liquid heat exchangers to abstract low grade heat from the atmosphere to warm up the evaporator section of the heat pump.

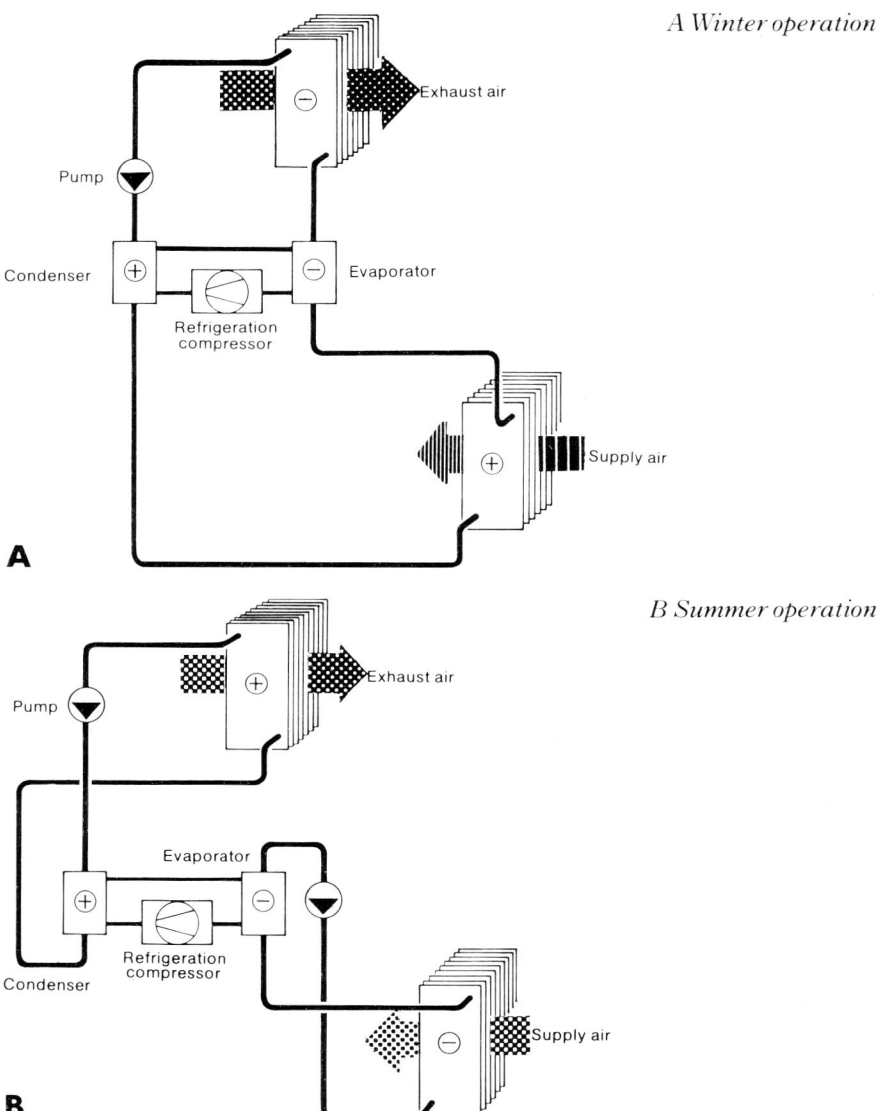

A Winter operation

B Summer operation

5.8 The Flakt Ecoterm heat and cold recovery system

INDUSTRIAL AND COMMERCIAL APPLICATIONS OF HEAT PUMPS

Industrial uses of heat pumps can be categorized into two types:
a Where it is necessary to install refrigeration equipment, as in food stores, ice cream factories, and the like. Under such circumstances the refrigerated area serves as the supplier of the low grade heat for the

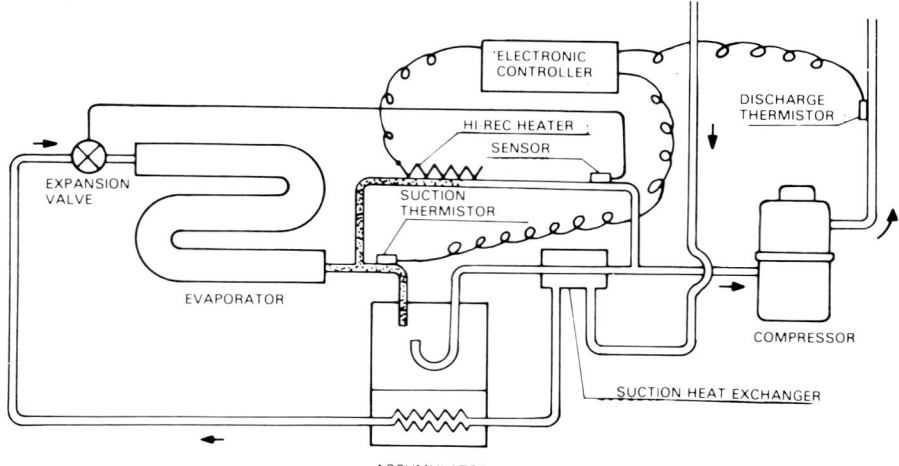

5.9 Central control of Daikin heat pumps

evaporation section of the heat pump. The condensation heat is used for space heating and process heat. The coefficient of performance of such a system can be raised by using other sources of waste heat to raise the temperature of the refrigerant after it leaves the refrigeration area. Such a system is infinitely preferable to the traditional system where waste heat is simply dissipated in a cooling tower.

b For all purposes where it is necessary to provide process heat at a higher temperature than that of the waste heat available. The heat pump in this case recovers thermal energy present in liquid and gaseous effluents and converts it into high grade heat for drying, evaporation, boiling and space heating. Although it is necessary to expend some high grade electric or other energy for driving the compressors, the advantageous COP obtained makes such a process an economic proposition.

Heat pumps for drying equipment

Heat pumps are widely used for this purpose. At the condensing side of the system air is heated sufficiently to drive moisture out of the damp materials to be dried. The moist air is then used as a heat sink on the evaporator section of the heat pump. Water condenses and drains off. The air which has had its moisture removed is then recycled to the warm part, where it is heated once more, to absorb further quantities of water from the medium to be dried.

In the various food, chemical and pharmaceutical industries of the United Kingdom it is necessary to remove about 25 million tonnes of water each year by evaporation. Drying by heat pump is far more efficient than conventional drying. In conventional drying one cannot recirculate the air which has carried out the job of drying the medium, as the air is saturated with moisture. It has to be released damp to the atmosphere. With heat pump drying, the water and sensible heat content of the damp air are both removed and the air is then recycled for further drying.

Assuming an overall dryer efficiency of 50 per cent roughly 5 GJ of thermal energy are required to remove 1 tonne of water in traditional drying equipment. The COP of a heat pump drying plant depends upon the temperature difference between the hot and cold section of the heat pump. The smaller this difference, the larger the COP.

Well-designed heat pump drying systems can evaporate a tonne of water for an energy expenditure of between 0.75 and 1.0 GJ, or between 15 and 20 per cent of the energy used by traditional drying plant. It is true that this energy is needed in the form of either electricity or another form of mechanical energy, instead of cheaper fossil fuel combustion. However, the difference in energy requirements is so pronounced that heat pump drying is still usually cheaper than conventional drying, particularly if one can employ cheap off-peak or total energy power.

Liquid effluents
Many liquid effluents not only carry away with them valuable sensible heat, but constitute an environmental hazard because they are warm. Discharging warm liquids into rivers is often forbidden because raising water temperatures by even a few degrees can kill off fish. Yet it is often inapplicable to remove the heat from such effluent flows by using simple heat exchangers as there is little need inside the works for low-grade heat.

Heat pumps are most useful in this context because they can cool down the warm effluents to a temperature at which they are harmless. At the same time, with the expenditure of quite a modest amount of prime energy, the thermal level of the extracted heat can be raised sufficiently for this heat to be useful for process heating.

A typical example of such equipment is the Westinghouse Templifier unit, which can extract heat from effluent fluids at temperatures between 27°C and 77°C, upgrading this to temperatures of between 60°C and 100°C respectively. Data given by the firm, stated in **table 5.3**, indicate the COP for a plant where the effluent enters the evaporator heat exchanger at 35°C and exits at 30°C.

Although the energy used for the Templifier has to be expensive electricity rather than cheap oil or gas, the actual running cost is lower. Comparisons were made between the annual costs of running a 1000 kW heat output system employing gas at a cost of $4.51 per GJ, oil at $5.05 per GJ and a Templifier unit with a COP of 4.4 employing electricity costing 4.708 cents per kWh. It was found that the actual annual fuel costs were as follows:

Gas $78 936
Oil $85 008
Templifier heat pump $42 768

Although the capital cost of a heat pump is higher than that of a simple gas or oil fired boiler, savings in fuel are so very considerable that the installation of heat pumps becomes a viable proposition.

Gas driven heat pumps

When one uses a prime mover to drive a heat pump one has to consider the considerable Carnot losses of the conversion of chemical energy into heat energy and then into mechanical energy. However, prime mover driven heat pumps are of value when one can use the waste heat from the gas engine for useful purposes. A gas engine converts approximately 31 per cent of the net calorific value of the fuel gas used into electrical energy. The remaining 69 per cent are contained in the following:

Exhaust gas: sensible heat and unburned gas   28%
Engine water jacket   30%
Oil cooler   3%
Miscellaneous heat losses   8%

The highest grade heat is from the exhaust, which is at an average temperature of 650°C. The exhaust cannot be cooled down below about 180°C because condensation and with it corrosion troubles may then arise. Assuming that a 75 kW(el) engine is used it becomes feasible to remove around 47 kW of the exhaust heat for such purposes as water heating or steam generation. Heat removal from the engine jacket and oil cooler normally takes place at about 110—120°C.

HEATING CYCLE

TABLE 5.3 COP VALUES OF HEAT PUMPS (WESTINGHOUSE)

Inlet temperature to evaporator section	35°C
Outlet temperature from evaporator section	30°C
Inlet temperature to condensing section	54°C
Outlet temperature from condensing section	66°C
Water flow through condensing section in litres/second	**COP (dl)**
9.2	4.21
19.7	4.28
29.4	4.48
46.9	4.66
68.3	4.90

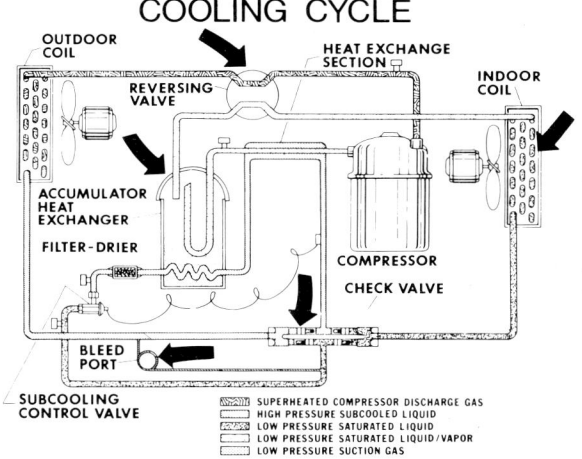

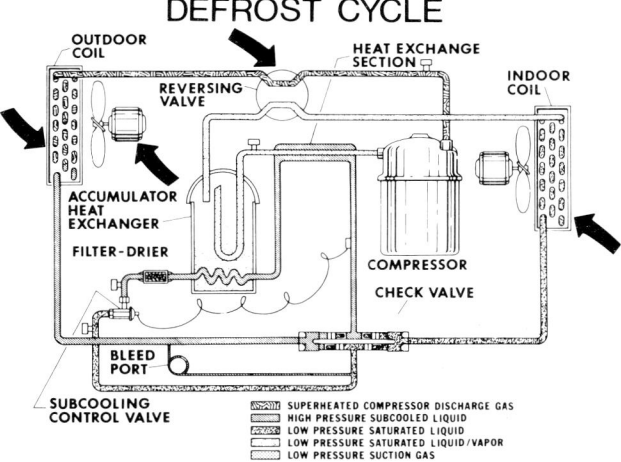

*5.10 Heating, cooling and defrost with typical heat pump. Courtesy
Westinghouse*

5.11 Gas engine driven heat pump for raising steam. Courtesy D. A. Reay, and International Research and Development Ltd

To sum up, it becomes possible to obtain 31 per cent of the chemical energy fed into a gas engine as useful work and another 49 per cent in the form of usable waste heat at a temperature above 110°C. 20 per cent of the chemical energy is not recoverable, for technical reasons.

Using refrigerant 114 it is possible to operate a heat pump with a COP of about 3.19, employing the mechanical energy supplied by the gas engine. It can therefore be calculated that the total heat supplied by the gas engine which is fed with 242 kW of fuel is:

$$\frac{242 \times 31 \times 3.19}{100} = 239.3 \text{ kW of heat pump heat and}$$

$$\frac{242 \times 49}{100} = 118.6 \text{ kW of usable waste heat}$$

giving a total of 357.9 kW of heat at a level of about 110—120°C. Assuming that the 242 kW gas supply was employed instead in a boiler plant with an overall efficiency of 80 per cent it would have delivered

$$\frac{242 \times 80}{100} = 193.6 \text{ kW of useful heat.}$$

The ratio of heat delivered by the gas engine/heat pump system as against the gas boiler is therefore

$$\frac{357.9}{193.6} = 1.849$$

In other words, the use of the gas engine/heat pump system provides a saving of approximately 46 per cent in primary fuel costs for the expense of the difference in capital cost between a simple gas-fired boiler installation and the gas engine/heat pump system.

Let us assume that the cost of gas is equal to 1.5¢ for 3600 kJ (1kWh) and that the difference in plant cost for a saving of $0.849 \times 242 = 205.5$ kW of gas supply equals \$70 000 (approximate cost figures applicable in 1981). If the plant is to be used 8000 hours per annum, then the saving of fuel is equal to:

$205.5 \times 0.015 \times 8000 = \$24 660$

This gives a pay-back period, excluding maintenance costs, of

$$\frac{70\,000}{24\,660} = 2.84 \text{ years}$$

Obviously detailed costing must be carried out in each individual case.

SELECTION CHART: HEAT PUMPS:

Type of equipment	Advantages	Disadvantages	Appropriate applications
Electrically driven	Easily installed High COP Cheap to buy Viable as small units	High cost of fuel	Domestic purposes Small industrial plant Combined with small air conditioning and refrigeration systems
Gas driven	Can use waste heat from gas engine Low fuel costs	Only feasible in larger units Low COP Expensive plant	Industrial application particularly where there are cooling requirements
Vapour recompression	Excellent efficiency	Only suitable for large installations	Energy recovery from sugar evaporators, sea-water desalination plants and similar
High temperature absorption cycle	Recovery of high grade waste heat	Expensive unit Only feasible for large plants	Industrial refrigeration plants or large scale air conditioning systems

References and further reading

1 E. Comatini and T. Kester, *Heat pumps*, Sithoff and Noordhoff, New York, 1976.

2 M. J. Collie, *Heat pump technology for saving energy*, Noyes, New York, 1979.

3 US Department of Energy, *Heat pump technology, a survey of technical development*, Washington, 1980.

4 N. W. Lord et alia, *Heat pump technology*, Ann Arbor Science, Michigan, 1980.

5 E. R. Ambrose, *Heat pumps and electric heating*, J. Wiley and Sons, New York, 1965.

6 D. A. Reay and D. B. A. Macmichael, *Heat pumps design and applications*, Pergamon Press, Oxford, 1979.

7 D. A. Heap, *Heat pumps*, Spon, London, 1979.

8 D. A. Reay, 'Industrial applications of heat pumps', *Civil and Municipal Engineer*, April 1980 pp 86–93, May 1980 pp 73–78.

9 Heating and Ventilating Contractors' Association, *Heat pumps for domestic applications*, London 1983.

10 Numerous communications from industrial organisations active in this field.

SELECTED COMPANIES INVOLVED IN MANUFACTURE OF PRODUCTS

UK and Europe

ACFT Louis Zhendre, 122 Avenue des Pyrénées, 33.140 Villenave D'Ornon/Pont De La Maye, France.

AEG Telefunken (UK) Ltd, 217 Bath Road, Slough, Berks SL1 4AW.

L'Air Conditionné Entreprises, 24 Boulevard de la République, cc78400 Chatou, France.

Airdale Air Conditioning Ltd, Clayton Wood Rise, West Park, Leeds LS16 6RF.

Aiax (UK) Ltd, 55 Bideford Avenue, Perivale, Greenford, Middx UB6 7PP.

Amcor Ltd, 98 Giborei Israel Street, PO Box 2850, Tel Aviv, Israel.

Andrews Industrial Equipment Ltd, Dudley Road, Wolverhampton WV2 3DB.

Bayford Tomlinson Ltd, 6 Lidgett Lane, Garforth, Leeds LS25 1EQ.

Biddle F. H. Ltd, Newtown Road, Nuneaton, Warks CV11 4HP.

Bitzer Kuhlmaschinenbau, Eschenbrunnlestrasse 15, D-7032 Sindelfingen, West Germany.

Conservatherm Ltd, Hereford House, Station Road, Billinghurst, West Sussex RH14 9SE.

Coronet Heat Pumps Ltd, Unit 2, The Causeway, Maldon, Essex.

Curwen and Newbery Division, KDG Industries Ltd, Fleming Way, Crawley, West Sussex RH10 2QE.

Daikin Europe NV, Zandvoordestraat 300, B 8400 Oostende, Belgium.

Danfoss A/S, Horsenden Lane South, Greenford, Middx UB6 7QE.
Delchi SpA, Via R. Sanzio 9, 20058 Villasanta, Italy.
Delta RA Ltd, Hollands Road, Haverhill Suffolk CB9 8PT.
Dunham Bush Ltd, Fitzherbert Road, Portsmouth PO6 1RR.
Eastwood Heating Developments Ltd, Portland Road, Shirebrook, Mansfield, Notts NG20 8TY.
Elstree Air Conditioning Ltd, Boreham Wood, Herts WD6 3AW.
Frimair SA, Zone Industrielle Sud, BP60 21600 Longuic, France.
Jaga NV, Verbindingsweg, B 3610 Diepenbeek, Holland.
Girdwood Halton (Air Conditioning) Ltd, The Street, Hockering, Dereham, Norfolk NR20 3HL.
Glynwed Integrated Services Ltd, Unit 15a, Shirley Trading Estate, Cranmore Drive, Cranmore Boulevard, Shirley, Solihull, West Midlands.
Hall Thermotank Products Ltd, Hythe Street, Dartford, Kent DA1 1BU.
International Research and Development Co. Ltd, Fossway, Newcastle-upon-Tyne NE6 2YD.
Marstair Ltd, Unit 4, Armytage Road, Industrial Estate, Wakefield Road, Brighouse, West Yorks.
Iain Miller Associates, Effie Road, London SW6 1EL.
Myson Industrial Group plc, Industrial Estate, Ongar, Essex CM5 9RE.
Officine di Seveso SpA, Via Orobia 3, 20139 Milano, Italy.
OY Nokia AB, PO Box 44, 01511 Vantaa 51, Finland.
Riello Condizionatori SpA, 37040 Bevilacqua, Verona, Via Roma 44, Italy.
Saunier Duval Ltd, 11 Decoy Road, Worthing, West Sussex BN14 8ND.
Schafer Werke GmbH, Pfannenberg, 5908 Neunkirchen 4 (D), West Germany.
Solaire Ltd, The Roofhouse, Windmill Lane, Avon Castle, Ringwood, Hants.
Stal-Levin Ltd, Yiewsley Heights, West Drayton, Middlesex UB7 7TA.
Stiebel Eltron Ltd, 25 Lyveden Road, Brackmills, Northampton NN4 0ED.
Tadiran Israel Electronics Industries Ltd, Electrical Appliance Division, POB 1758, Holon, Israel.
Technibel, BP262-RD28 Reyrieux, 01600 Trevoux, France.
Trace Heat Pumps Ltd, Industrial Estate, West Witham, Essex CM8 3TQ.
tt Coils ApS, Svanningeved 2, 9220 Aalborg Øst, Denmark.
Walker Air Conditioning Ltd, 30 Sloane Street, London SW1X 9NJ.
Westwarm, Unit 6, Hither Green, Clevedon, Avon BS21 6XT.
Willison Controls Ltd, Dallas Road, Bedford MK42 9ES.

United States
ARE Manufacturing Co., PO Box 113, Frederick, MD 21701.
American Heat Pump Mfg Inc., 8001 Franklin Farms Drive, Richmond, VA 23288.
Arkla Industries Inc., PO Box 534, Evansville, IN 47704.
Bard Mfg Co., PO Box 607, Bryan, OH 43506.
BDP Co. Div. of United Technologies Corp., 7310 W. Morris Street, PO Box 70, Indianapolis, IN 46206.
Bristol Compressors Inc., 649 Industrial Park Road, Bristol, VA 24201.
California Heat Pump Co., 2314 Michigan Avenue, Santa Monica, CA 90404.
Coleman Co. Inc., 250 N. St. Francis Street, Wichita, KS 67201.

Command-Aire Corp., 3321 Speight, PO Box 7916, Waco, TX 76710.
Drake Industries Inc., PO Box 8186, Fort Lauderdale, FL 33310.
Duo-Therm Division, 509 S. Poplar Street, Lagrange, IN 46761.
Energy Conservation Systems Inc., 1775 Central Fla. Parkway, Orlando, FL 32809.
Energy Management Engineering Inc., 4241 Hogue Road, Evansville, IN 47712.
FHP Manufacturing Div., 610 S.W. 12th Avenue, Pompano Beach, FL 33060.
Friedrich Air Cond. and Refn. Co., 4200 N. Pan Am, PO Box 1540, San Antonio, TX 78295.
General Electric Co., Central Air Cond. Dept., Troup Highway, Tyler, TX 75711
Heil Heating and Cooling Products, 635 Thompson Lane, PO Box 40566, Nashville TN 37204.
Intertherm Inc., 10820 Sunset Office Drive, St. Louis, MO 63127.
Introdel Inc., 246 Woodwork Lane, Palatine, IL 60067.
Jacobsen Energy Industries Inc., 651 Vernon Way. El Cajon, CA 92020.
Melcor Materials Elect. Prods. Corp., 990 Spruce Street, Trenton, NJ 08648.
Modine Mfg Co., 1500 DeKoven Avenue, Racine, WI 53401.
Natural Energy Systems Inc., 3407 N. Ridge Avenue, Arlington Heights, IL 60004.
Northrup Inc., 302 Nichols Drive, Hutchins, TX 75141.
Phoenix Envirotech, 651 Vernon Way, El Cajon, CA 92020.
Singer Climate Control Products, 62 Columbus Street, Auburn, NY 13021.
Solar Energy Research Corp. 10075 E. County Line Road, Longmont, CO 80501.
Square D. Co. Sun Dial Plant, 800 E. Kearney, Mesquite, TX 75149.
Thermodyne Inc., 506 Jasper Street, PO Box 1007, West Columbia, SC 29171.
Turbo Refrigeration Co., 1515 Shady Oaks Drive, Denton, TX 76201.
Watercotte Corp., 3407 N. Ridge Avenue, Arlington Heights, IL 60004.
Weatherking Inc., 4501 E. Colonial Drive, Orlando, FL 32814.
Westcorp Inc., 15 Stevens Street, Andover, MA 01810.

6 Energy storage

Whenever we produce heat or power by non-traditional methods (ie those which do not involve the combustion of fossil fuels) we face the problem of matching demand with supply. Nuclear power stations operate at a virtually steady output day and night, summer and winter and cannot easily be switched off or turned down as can be done with coal and oil-fired stations. Solar power is at a maximum during midday in summer: at periods when demand for electricity is quite low and heat demand is virtually zero. Tidal power, wave power, wind power and the like also peak at varying times which do not coincide with periods of maximum energy demand.

Increasingly therefore, it will become necessary to build systems of energy storage into our energy economy. These aim to store heat or electric power during periods of high supply but low demand, such as at night and during the summer, and release this energy once more at periods of maximum demand.

The main methods of energy storage used are the following:

Sensible heat
Latent heat
Chemical
Electrochemical
Compressed air
Flywheels
Pumped water
Steam

Sensible heat storage

Such heat storage is basically for low-grade heat such as solar energy or waste heat from power generation and industrial processes. Water as a heat storage medium has excellent specific heat and is both cheap and chemically stable. If it is used above 100°C the system has to be pressurized, which adds enormously to costs. In any case the limitation of water is its critical point, namely 374°C. There are a number of heat-resistant oils on the market, such as Caloria HT43 or Therminol. They can readily be used without pressurization at temperatures as low as −10°C and up to 320°C. While water has an average specific heat of about 4.19 kJ/kg K, most of these oils have specific heats of only about 2.3 kJ/kg K. They also have the disadvantage of liability to high

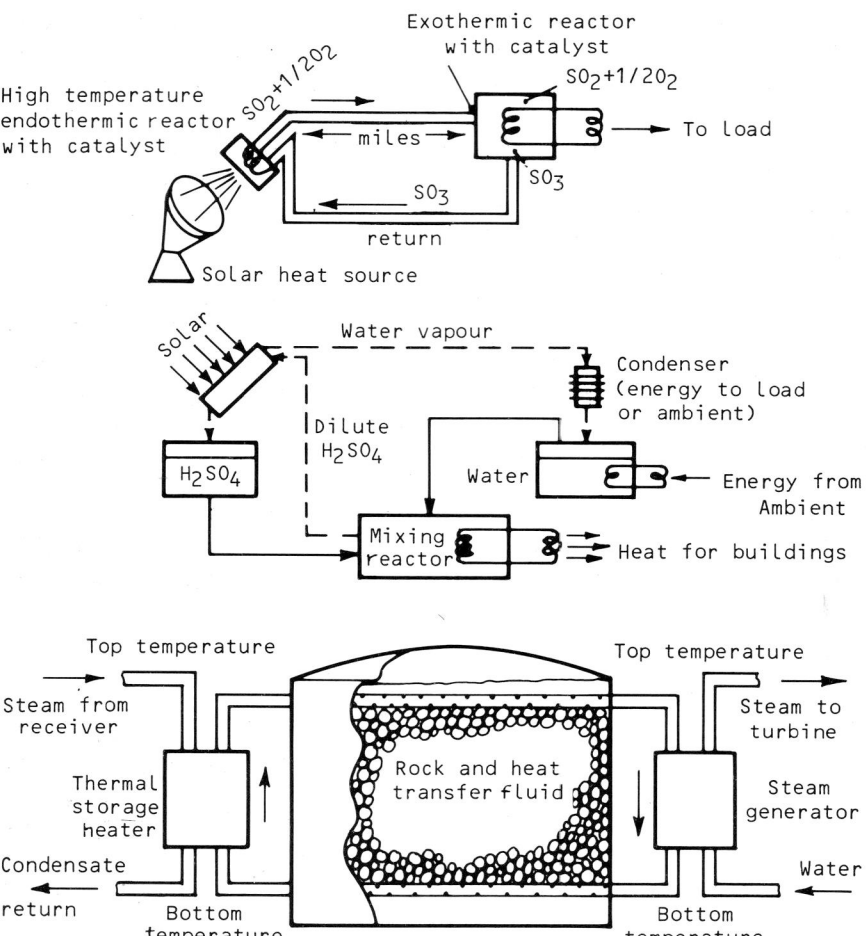

6.1 *Various heat storage systems. Source: Wyman et al, Solar Research Institute, Golden County, USA*

temperature cracking, polymerization and the formation of volatile products.

Molten salts can also be used for high temperature heat storage. Specific heats average 1.5 kJ/kg K and other disadvantages are solidification temperatures of a minimum of 150°C, as well as a corrosive nature.

Molten metals such as liquid sodium can be used in the unpressurized state at temperatures of up to 760°C. The specific heat is only 1.3 kJ/kg K and there are also considerable handling problems.

Solid rock heat storage
Many rocks are used as heat stores, particularly magnesium and

aluminium silicates. The higher the magnesium and aluminium oxide contents, the better the thermal energy capacity. Rock stores of this type are commonly used with water as a heat exchange medium and, in the absence of pressurization, the peak storage temperature is then 80°C. When air is used as a heat exchange medium the peak temperature of such a rock store can be held at 300°C. Under such circumstances the storage capacity rises to as much as 1 GJ/m^3. Practical uses for heat storage systems of this type are twofold:

1 One can use large-scale thermal stores in conjunction with district heating networks. All district heating systems store thermal energy in the circulating hot water and also in huge, well insulated accumulator systems. The water is heated up at night by operating pass-out turbines at maximum steam bleed and is then used at peak daytime heat and power consumption periods. Modern developments are additional heat storage systems for the utilization of waste nuclear heat, waste industrial heat and solar energy for district heating.

2 As the heat level of sensible heat energy stores is often below the level of heat actually required, one makes use of heat pumps to upgrade this heat. Although heat pumps use precious electric energy, using lukewarm water as a heat sink they can produce about 3–3.5 times their

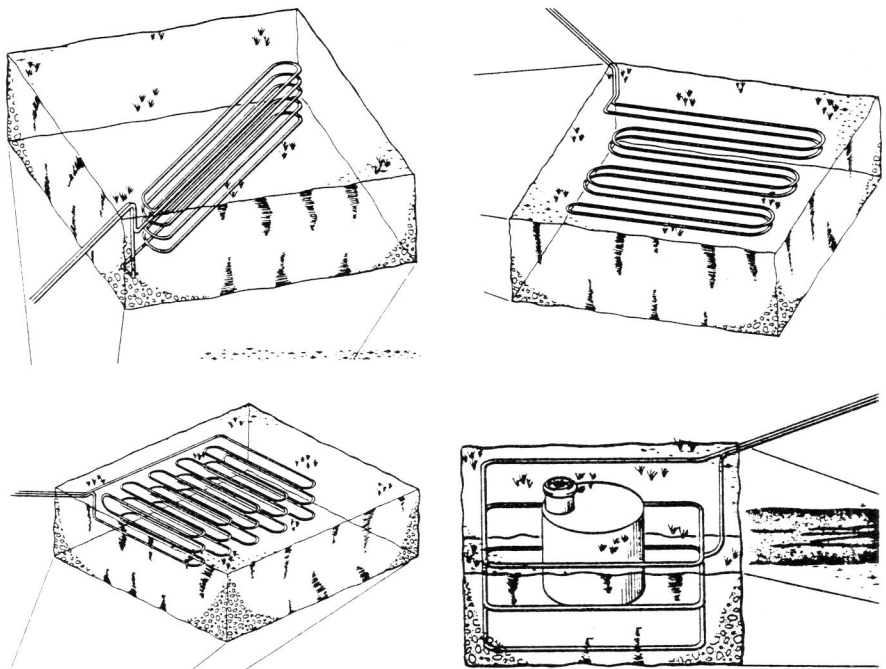

6.2 *Systems of underground low-level heat storage. Courtesy P. D. Metz, Brookhaven National Laboratory*

Sendout (150°C, 300°F)

Hot water transmission loop

Return (60°C, 140°F)

Heat exchanger

HEAT STORAGE WELL DOUBLET

Confining Cap

·Hot·   Cold   ‾Warm‾

Confining Bottom

Flow direction: ▷ Storing heat   ▶ Withdrawing heat

6.3 Heat storage well. Courtesy GEC

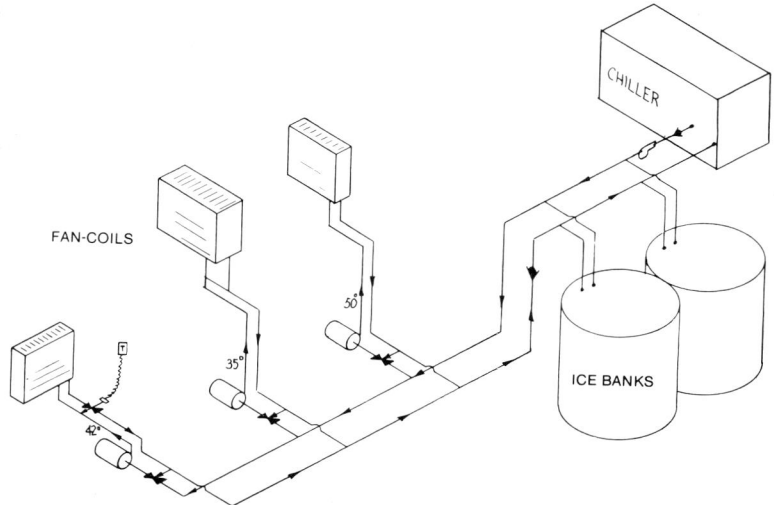

CHILLER

FAN-COILS

50°

35°

42°

ICE BANKS

6.4 Refrigeration energy storage. Courtesy Calmac, Englewood, NJ
Charge cycle: anti-freeze, cooled by standard chiller-type refrigeration
equipment, is circulated into the ice bank modules. The anti-freeze extracts
heat through a specially designed mat-type heat exchanger, until
eventually all the water in the tank is frozen solid.
Discharge cycle: stored cooling power is drawn from the modules through
same heat exchanger and circulated to the load in the conventional way

power consumption in the form of high temperature hot water. In addition, they yield valuable refrigeration energy.

Latent heat storage

The heat change which takes place when a substance passes from one phase to another is called latent heat and is always much higher than the purely sensible heat storage capacity of the medium by virtue of its specific heat. When water turns into steam, latent heats of the order of 2MJ/kg are involved.

Most practical systems of phase change energy storage involve solutions of salts in water. Problems are generally threefold:

1 Supercooling may take place, rather than crystallization with heat emission. This can be avoided partially by adding small crystals as nucleating agents.

2 It is difficult to build a heat exchanger able to deal with an agglomeration of varying sizes of crystals which float in the liquid.

3 The system is not completely reversible.

Some systems which use either Na_2SO_4 . 10 H_2O or $CaCl_2$. 6 H_2O crystals as heat store employ a heat exchange oil. This is pumped in at the bottom and rises in globules through the fluid without mixing. Other promising latent heat storage reactions are those of intercrystalline changes. Many of these take place at quite high temperatures. **Table 6.1** lists those which have been used.

TABLE 6.1 INTERCRYSTALLINE LATENT HEAT STORAGE MEDIA

System	Temperature of turn in °C
$KF-NaF-MgF_2$	685°C
$Na_2CO_3-K_2CO_3$	710°C
NaOH	318°C
$NaNO_3$	307°C
$KCl-Na_2CO_3$	588°C
$CaCl_2-NaCl$	500°C

The temperature of turn is the temperature at which heat has to be pumped in, in order for it to be stored. Once the heat level of the storage medium drops appreciably below the temperature of turn, this heat is given off once more.

Aquifer thermal energy storage

Almost a third of the annual energy requirements in temperate climatic zones is for winter space heating. The thermal grading required for this is quite moderate and for this reason it is feasible to consider such

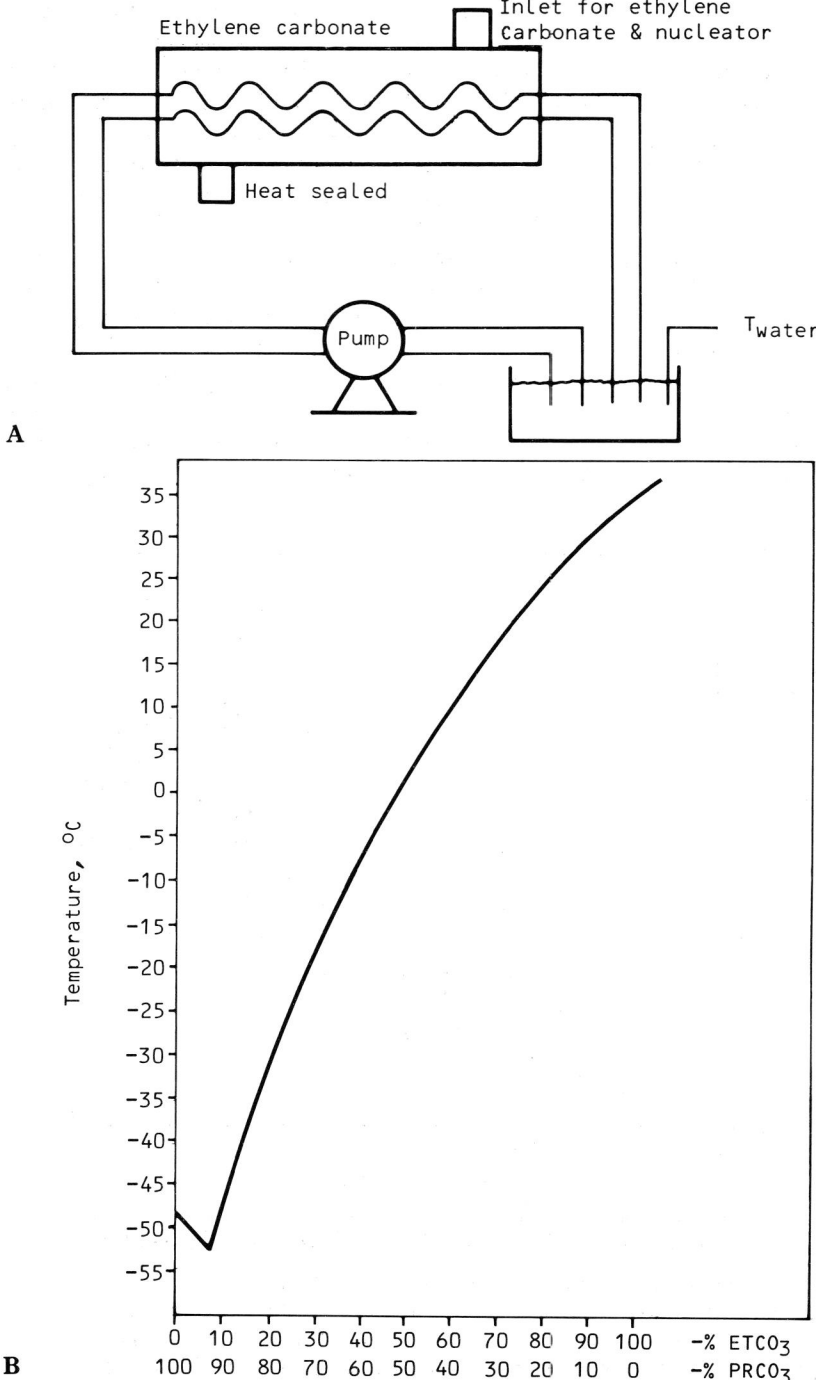

sources as waste heat from power stations, solar energy or other waste heat.

Geothermal heat is already being used for district heating projects in Iceland, Japan, the Soviet Union, Hungary and France. Artificially heated underground water resources, so-called aquifers, are now being studied in many countries. Typical of such an aquifer experiment is the one carried out at Auburn University, Syracuse, NY, USA. An aquifer is maintained there at 20°C by the injection of water at 55°C. Almost 58 000 m³ can be injected over a period of 80 days, stored for 51 days and then withdrawn for use. The energy recovery ratio is 65 per cent.

Aquifers, when reasonably well insulated by natural rocks and sand, would give the opportunity of storing waste nuclear and solar heat during the summer months, making possible its recovery and use during the winter. The main problems so far encountered are leakages of the aquifer-stored hot water, thermal dispersion and corrosion. The last problem is caused by the dissolution of salts within the water layer.

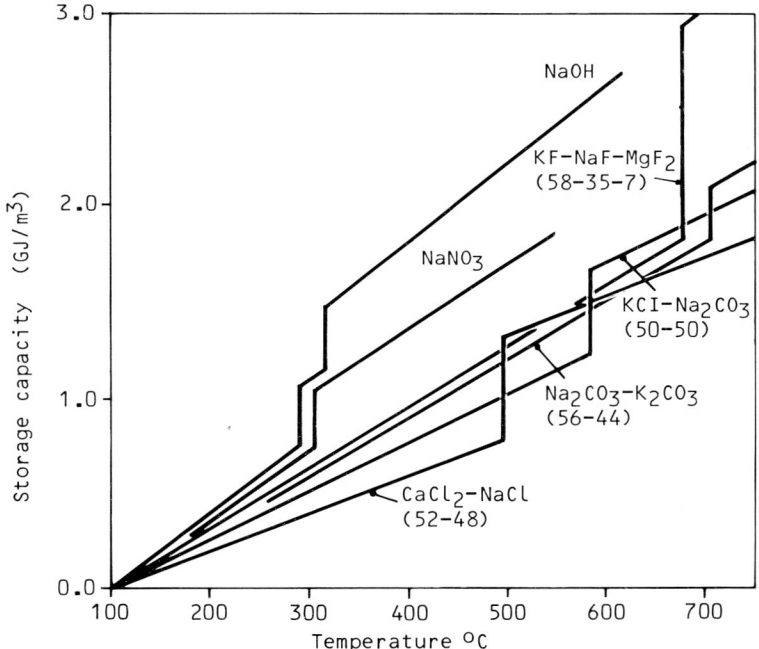

6.6 (above) Heat storage capacity of selected latent heat storage materials.
  Source: Wyman et alia Solar Energy Research Institute, Golden Co.

6.5 (left) Ethylene carbonate as a heat storage medium.
  A Schematic representation of test rig for a closed loop water system
  B Freezing points of ethylene carbonate-propylene carbonate mixtures
  Both diagrams courtesy Dr Raj Gopal, Johnson Controls, Milwaukee

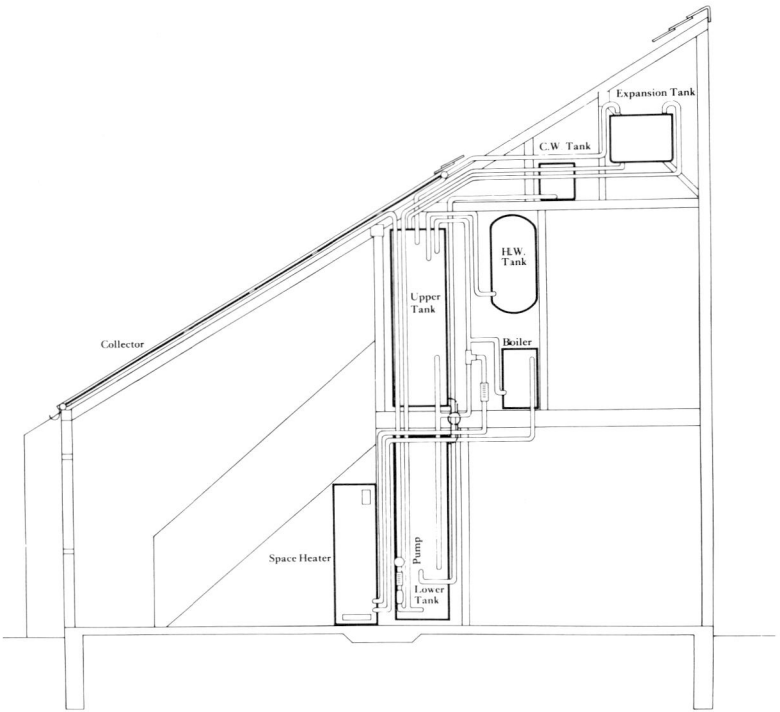

6.7 Solar collectors providing domestic hot water and space heating. Courtesy Milton Keynes Development Corporation

6.8 Storage of solar energy in Denmark. Courtesy Dr O. Rothmann, Riso National Laboratory, Denmark

Energy storage in conjunction with solar energy utilization
Several practical methods of energy storage have been developed with
systems of using solar heat. Solar collectors are used to transfer solar
energy to a stream of flowing water which heats up a well-insulated
underground heat storage reservoir. The heat can either be utilized
directly, when needed for space heating, or it can be stepped up by a
heat pump to produce higher-grade energy.

Typical of such a plant is a prototype described by S. W. Yuan and M.
M. Majdi. Two 54.8 m^2 Owens Illinois Sunpak solar collectors abstract
solar energy, which is used to heat up water. An 82 m^3 earth heat storage system is used which is heated up by four 50 mm diameter poly-
butylene tubes, each of which is 40 m in length.

Swedish solar energy enthusiasts have visualized the heating of circulating water by solar energy and its collection in large, highly insulated
underground water tanks. During winter this water is then circulated to
supply both central heating and domestic hot water to about 100 dwelling units per system. This would have a solar collector area of 1000—
2000 m^2, and a hot water storage capacity of 1000—2000 m^3. It is estimated that the storage capacity needed per dwelling is about 225 kWh
of heat in the form of hot water.

Chemical storage

Many chemical reactions proceed forward with absorption of thermal
energy, ie they are *endothermic*. According to the first law of thermodynamics, energy can neither be created nor destroyed. The entire heat
which has been absorbed by the endothermic reaction is thus released
in the reverse reaction which is *exothermic*, ie gives out heat. In all
cases there is, however, a loss in *heat grade*, ie the heat given out is at a
lower temperature than the heat absorbed. Heat changes are commonly
given in terms of kilojoules per gramme mole of reacting materials (the
molecular weight of the substances in grammes).

Typical of such reactions is that of sulphuric acid with water to form
the hydrated state. This can be expressed as:

$$H_2SO_4 + n\,H_2O \rightarrow H_2SO_4.n\,H_2O \qquad \Delta H = -23 \text{ kJ/mole}$$

As the molecular weight of sulphuric acid is 98 g, each kg of concentrated sulphuric acid yields about 235 kJ of energy when diluted to
infinity.

This system has been used as a practical means of energy storage. At
periods when excessive heat is produced, waste thermal energy is used
to drive off water vapour from dilute sulphuric acid and obtain concentrated acid. When the heat is needed, water is simply added to the acid
and the reaction, which is now exothermic, releases heat once more.

In view of the dangerous and corrosive nature of sulphuric acid, the
plant is entirely enclosed and constructed from corrosion resistant
materials. The main problem, however, is that the system is not

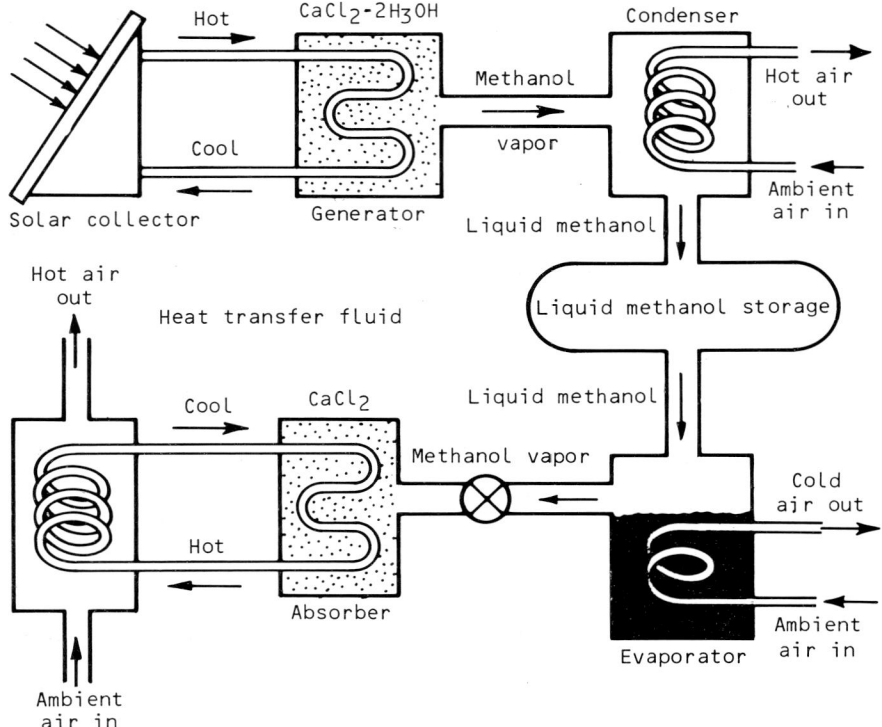

Note: Generator and absorber beds are exchanged on alternate cycles

6.9 Heat pump using methanol/calcium chloride reaction. Courtesy Offenhartz, EIC Laboratories, Newton, Mass

entirely thermodynamically reversible. The temperature level at which it is necessary to drive off the water from the dilute sulphuric acid is far higher than the temperature level of the heat produced when sulphuric acid and water recombine. There is also some waste of thermal energy as it is necessary to abstract latent heat from the water vapour driven off during the evaporation process, to condense it.

Other thermochemical reactions used for energy storage
a Decomposition of metallic hydroxides into metallic oxides and water.
b Decomposition of metallic carbonates into oxides and carbon dioxide.
c Decomposition of higher oxides into lower oxides and oxygen.
d Decomposition of higher ammonium salts into lower ammonium salts and ammonia.
f Decomposition of higher alcoholates into lower alcoholates and alcohol.

All these reactions are reversible and proceed forward as endothermic reactions, ie they absorb thermal energy. Equally, once the temperature of the system falls below a certain value, the heat which is stored in the system during the decomposition reaction is once more given off as the reaction processes are reversed.

Chemical heat pipes

These are devices by which energy can be pumped from the heat source, which is normally a nuclear reactor, to the heat consumer, using reversible reactions which are endothermic in one direction and exothermic in the opposite one. Such heat pipes also act as energy storage devices because gases, unlike electricity, can be stored.

A very common reaction is the following one:

$$CH_4 + H_2O \leftrightarrows CO + 3 H_2 \qquad \Delta H = 221.8 \text{ kJ/mole}$$

A mix of methane and steam is heated up from waste heat of the nuclear reactor and is converted into hydrogen and carbon monoxide by the absorption of 221.8 kJ of thermal energy per mole of mix. This is then pumped to the consumer area and stored until required. The reverse reaction proceeds by the use of a catalyst, and the 221.8 kJ/mole of heat are released and used for domestic or industrial purposes.

The hydrogen economy

Perhaps the most widely useful way of storing energy is in the form of hydrogen.

The reaction: $H_2 + \frac{1}{2}O_2 \rightarrow H_2O$

is very strongly exothermic, being accompanied by the evolution of 286.3 kJ of energy per mole of hydrogen burned. One of the great advantages of this reaction is the fact that, unlike the combustion of hydrocarbons, no obnoxious or polluting byproducts are formed.

Hydrogen can be formed from water by two methods, electrolysis, or indirect thermal cracking.

1 Electrolysis

Hydrogen is obtained as a byproduct of the electrolytic manufacture of sodium hydroxide and chlorine from sea water, but the current efficiency of this is only about 75 per cent. When one multiplies such a low efficiency by the even lower efficiency of obtaining electricity from fossil fuels, it becomes clear that electrolytic methods of obtaining hydrogen from oil or coal cannot be higher than:

$$\frac{35 \times 75}{100} = 26.3 \text{ per cent}$$

The use of hydrogen is therefore uneconomic if the only way of making electricity is by the combustion of fossil fuels. Hydrogen for the storage of energy, even using electrolytic means, will be of enormous value once oil reserves have dwindled to such an extent that combustion of oil and oil products has become prohibitive in cost, and nuclear power is

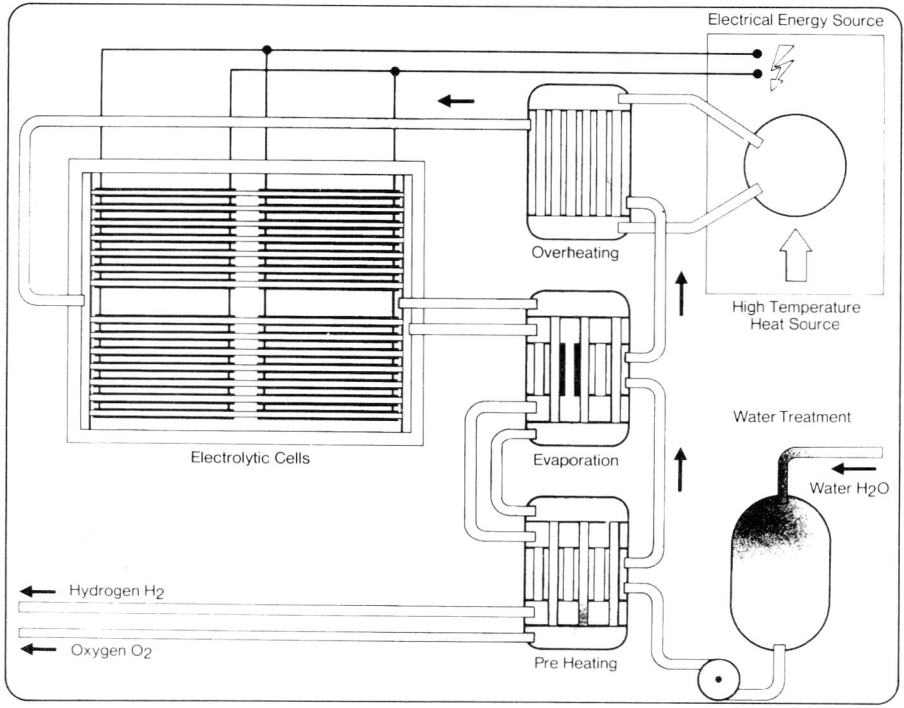

6.10 High temperature vapour phase electrolysis of steam. Courtesy Dornier GmbH

used almost exclusively to produce electricity. It has been found that electrolytic decomposition of water into hydrogen and oxygen proceeds at a much higher efficiency if temperatures are kept high.

Some electrolytic techniques use temperatures of up to 1000°C employing doped zirconia as electrolytes. Steam, which at such high temperatures already contains 2 per cent of free hydrogen, is then electrolysed.

A method used by the Dornier GmbH company of Friedrichshafen employs a ceramic electrolyte of mixed ZrO_2/Y_2O_3 at 800°C. The anode is the metallic oxide mix, while the cathode is of a nickel-cobalt alloy.

At anode: $H_2O + 2e \rightarrow O^{2-} + H_2$

At cathode: $O^{2-} - 2e \rightarrow \frac{1}{2}O_2$

A current efficiency of over 95 per cent is claimed.

Electrolytic methods of hydrogen production have been fully developed already but suffer from the following two disadvantages:

1 Their efficiency of energy utilization cannot exceed the efficiency of electricity production, which is only 30–35 per cent.

2 While small-scale electrolytic plants are economical, it becomes difficult to achieve economy in size when building large plants.

Indirect thermal cracking

It is extremely difficult to split water into its constituent elements of hydrogen and oxygen by heat alone. Even at 3000°C only about 5 per cent dissociation is achieved. It is therefore essential to use indirect thermal cracking techniques. Typical of these is the iodine/sulphur dioxide method, which is given here:

1   $2 H_2O + I_2 + SO_2 \rightarrow H_2SO_4 + 2HI$ (room temperature)
2   $H_2SO_4 \rightarrow H_2O + SO_2 + \frac{1}{2} O_2$ (800°C)
3   $2 HI \rightarrow H_2 + I_2$ (600°C)

The hydrogen and oxygen produced are separated off and constitute the product, while the sulphur dioxide and iodine are recycled to react with more water. At present the efficiency of this process is no more than 31 per cent and is therefore little better than some of the electrolytic techniques.

Another drawback which this method shares with other indirect water cracking techniques is the high temperature at which at least one of the reactions proceeds. In the above case the second reaction proceeds only to any appreciable extent at temperatures of 800°C and higher. Most nuclear reactors cannot achieve such high operational temperatures.

How hydrogen is used

Hydrogen can be stored in two ways:

1 As *liquid hydrogen* in highly insulated vessels which are kept at −258°C. In this form hydrogen becomes a very useful fuel for aircraft and for driving industrial gas turbines. Experimental aircraft which run on liquid hydrogen have already been designed by the Lockheed Aircraft Corporation.

2 In the form of metastable metallic hydrides such as:

LiH, $ZnMn_2 H_{3.3}$, MgH_2, Mg_2NiH_4, VH_2, $FeTiH_{1.9}$, $LaNi_5H_7$ and $Ni_5H_{6.5}$

Hydrogen can be stored in slightly pressurized tanks containing these compounds with considerable safety. These techniques are the ones most likely to be used for the fuelling of motor vehicles.

Other likely uses for hydrogen will be use in fuel cells to produce electricity at peak periods from hydrogen made at off-peak periods (summer or night) from excess nuclear heat or electricity, and the manufacture of a more conventional motor fuel by the following reactions.

1   CO_2 (present in air) $+ Ca(OH)_2 \rightarrow CaCO_3 + H_2O$
2    $CaCO_3 + H_2SO_4 \rightarrow CaSO_4 + CO_2$ (gas)
3    $CO_2 + H_2 \rightarrow CO + H_2O$
4    $CO + 2H_2 \rightarrow CH_3OH$ (methanol)

No fossil fuel whatsoever is needed to produce the methanol, which can be used like any traditional hydrocarbon liquid in internal combustion engines. The carbon dioxide is removed from the air by a liquid/gas scrubbing method.

Electro-chemical storage

It is notoriously difficult to store electrical energy. There is little chance of using conventional battery design to store more than an infinitesimal amount of the high voltage AC power which comes from either conventional or nuclear power plants in rechargable batteries. The problems are the following:

a The amount of energy which can be stored per unit weight is small.

b The power density per unit weight is low.

c All reversible batteries can withstand only a limited number of charge/discharge cycles, after which they have to be discarded.

   Table 6.2 gives the very best achievements in this field which have so far been obtained with a number of commercial batteries.

TABLE 6.2 BATTERY PERFORMANCE AND LIFE

Type of battery	Energy density Wh/kg	Power density W/kg	Maximum cycles before battery write-off
Lead/sulphuric acid	35	175	2000
Nickel/cadmium	45	600	3000
Nickel/iron	45	90	5000
Nickel/zinc	70	200	300
Silver/zinc	150	400	2000

From the point of view of cost the lead/sulphuric acid cell is still by far the best. It is the type used in most experimental electric cars designed in recent years.

   As can be seen, the optimum energy storage of a lead/acid battery is only 126 kJ per kg, admittedly 100 per cent useful electrical energy. The storage capacity of petrol is about 35 000 kJ per kg gross. Even when calculated on the basis that only about 18 per cent of this can be used as mechanical energy, this still works out at 6300 kJ per kg or exactly 50 times as much as the lead/acid battery. Looking towards the future, some of the alkali/metal high temperature batteries should have some promise. Claims have been made that high temperature lithium-aluminium/FeS_2 batteries should have a storage capacity of up to 200 Wh/kg (720 kJ/kg). This is, of course, still only 11.5 per cent of the storage capacity of petrol in a normal petrol tank. Still better storage capacities are promised for the chlorine hydrate/zinc system which has

*6.11 This experimental bus built by the Billings Energy Corporation uses liquid
hydrogen as fuel*

been used with the Vega experimental motorcar, which was propelled
by a 200 V, 100 A power plant. These batteries have a theoretical storage capacity of 830 Wh/kg (3000 kJ/kg). Furthermore it has already
been shown that it is possible to build storage modules of 45 kWh of
such systems. Once various difficulties inherent with this system have
been ironed out, it becomes a strong contender for use with road vehicles and as a storage system for large power stations. A facility is now
being built at Princeton, New Jersey, USA for the construction of a
100 MWh Zn/Cl_2 system. Fuel cells are also being developed to act as
large scale energy storage devices.

Compressed air storage

It is envisaged that a normal power station should include a compressor
system. During off-peak operation ambient air is compressed by an
axial flow compressor, intercooled and boosted to a pressure of approximately 70 bars by a high speed centrifugal blower. Any heat produced
during the compression stage has to be taken off by normal cooling
devices. The air thus compressed to 70 bars is passed to underground
storage. Leached-out salt domes would be ideal for such purposes. The
pressure of the soil would serve to resist the appreciable gas pressure.
At peak demand periods, air is led from the underground compressed
air storage cavern through a control valve, where the pressure is throttled down to 43 bars at full load. After being heated, the compressed air

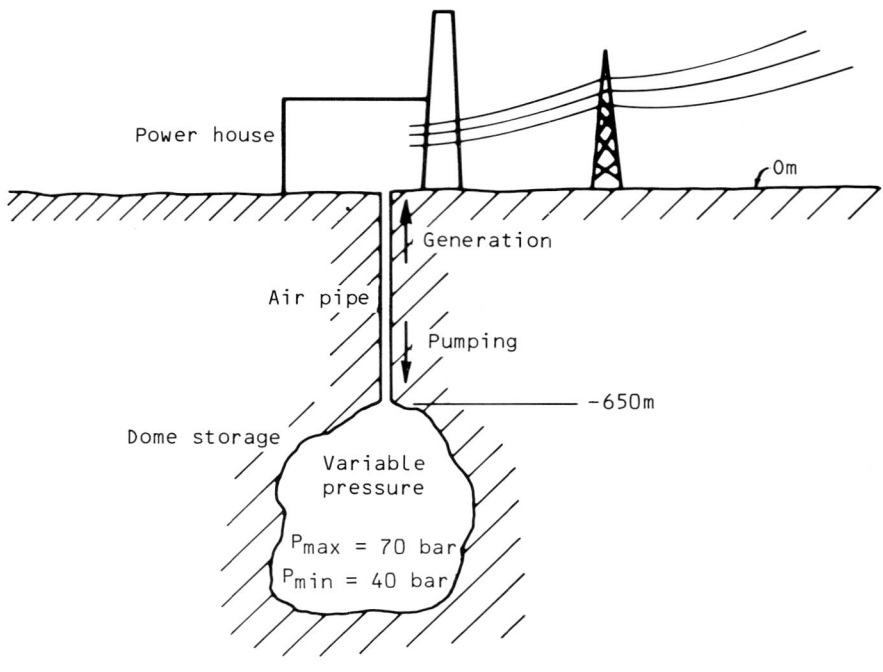

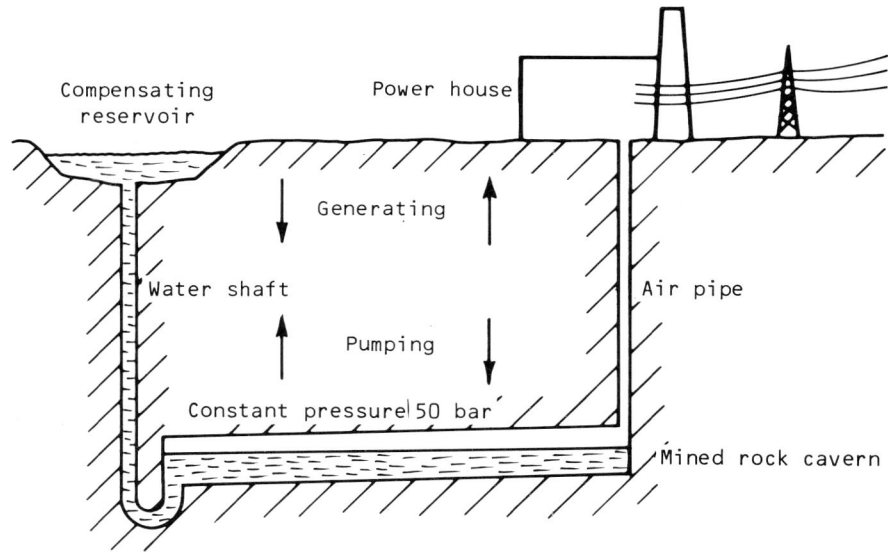

6.12 Compressed air energy storage systems   B Compensated pressure   A Sliding pressure. Courtesy Societe Electrique de l'Our, Luxembourg

can be used to drive gas turbines, effectively using up the energy stored from the off-peak period.

The Brown Boveri company has developed the ASSET (air storage system of energy transfer) plant for this purpose. The first plant of this type was ordered by NWK (Nordwestdeutsche Kraftwerke) of Hamburg in June 1974.

In 1980 the various powers station operated by this utility had the following capacities:

Fossil-fuelled steam plants	3000 MW
Nuclear plants	1210 MW
Gas turbines	305 MW
Total	4515 MW

Two caverns were formed from leached-out salt domes giving a combined compressed air storage volume of 300 000 m^3. This is enough for two hours of full peak load for the entire NWK system. The caverns and the power plant were completed in 1977 at Huntorf, some 100 km from the City of Hamburg.

No electric power is needed to start the unit. The compressed air is merely heated to 540°C in the high pressure combustion chamber of the gas turbine plant and is exhausted to the gas turbine. As soon as the gas turbine revolves at the specified speed of 3000 rpm, the generator is connected to it by a so-called synchro-self-shifting clutch. The total start-up time from zero to full load is normally 11 minutes, but in an emergency this can be reduced to a mere 6 minutes. All operations are carried out by remote control from Hamburg.

Due to the fact that heat supplied to the compressed air has to be taken off before compressed air storage and needs to be re-supplied before the air can drive the gas turbine, the system has a poor thermo-dynamic efficiency, estimated as being somewhat less than 46 per cent. This method is likely to be particularly valuable in systems where an appreciable part of the power load is carried by nuclear stations and where suitable spent salt caverns make it easy to build the compressed gas reservoirs.

Flywheel storage

To facilitate mechanical energy storage by flywheel it is necessary to develop low energy-loss and long-life bearings as well as suitable materials for the flywheels themselves. Materials which have been tested have been steel wire, vinyl-impregnated fibreglass, carbon fibre and other materials. SNI Aerospatiale have developed a series of industrial kinetic energy flywheel storage systems using magnetic bearings. A typical experimental flywheel capable of storing up to 3 kW of energy consists of a series of discs 400 mm in diameter and 200 mm deep with a spinning mass of about 240 kg. The system is suspended on six magnetic bearings which carry the weight when the unit is in operation, and two roller bearings which are used during the process of getting the unit to working speed. The flywheel is intended to be operated at speeds between 7 500 and 15 000 rpm.

A larger unit designed for a maximum power storage of 10 kW and up to a total of 200 W/h is intended for maximum speeds of 24 000 rpm. Both the flywheel and the ball bearings which support it are kept under vacuum.

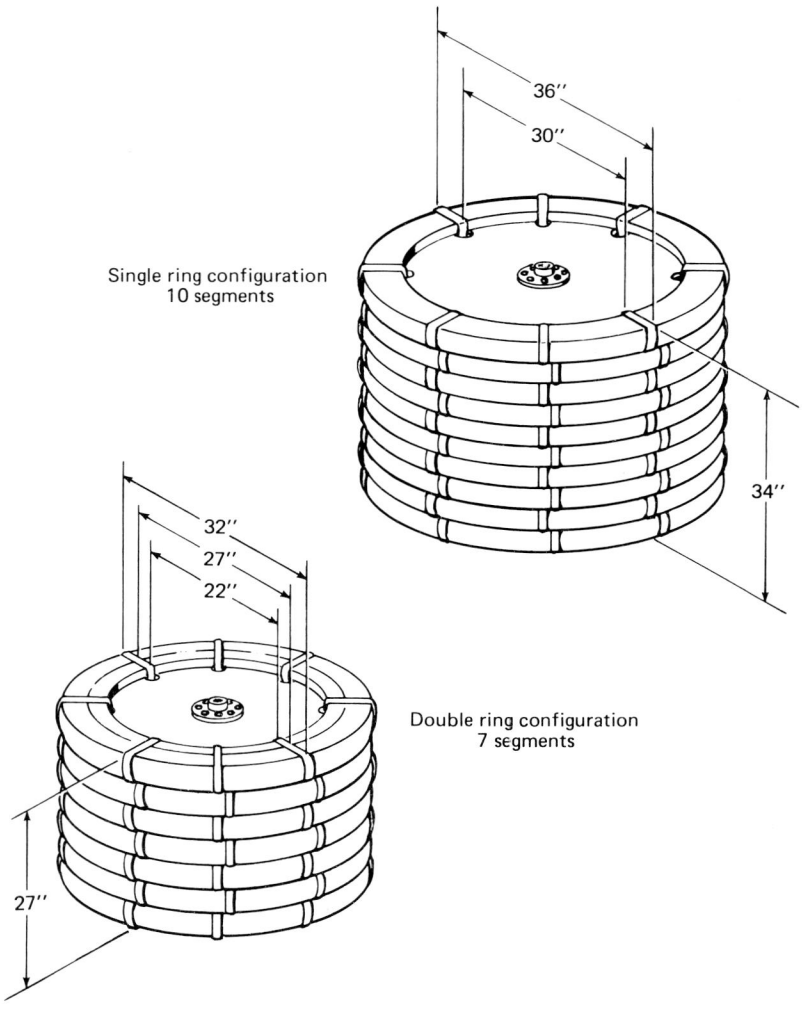

6.13 A 30 kWh energy storage flywheel, and a modular array of flywheels. Courtesy Johns Hopkins University

American experiments have been carried out with flywheels wound from Metglas, an amorphous metal ribbon made by the Allied Chemical Corporation. The ribbon is usually 12.5 mm wide and 0.05 mm thick, and can be wound easily and cheaply to form flywheels of the necessary strength and dimensional accuracy. Feasibility studies have shown that it would be possible to build economic single-family flywheel energy storage units capable of storing up to 30 kWh of power.

The idea would be to draw off cheap electricity during off-peak

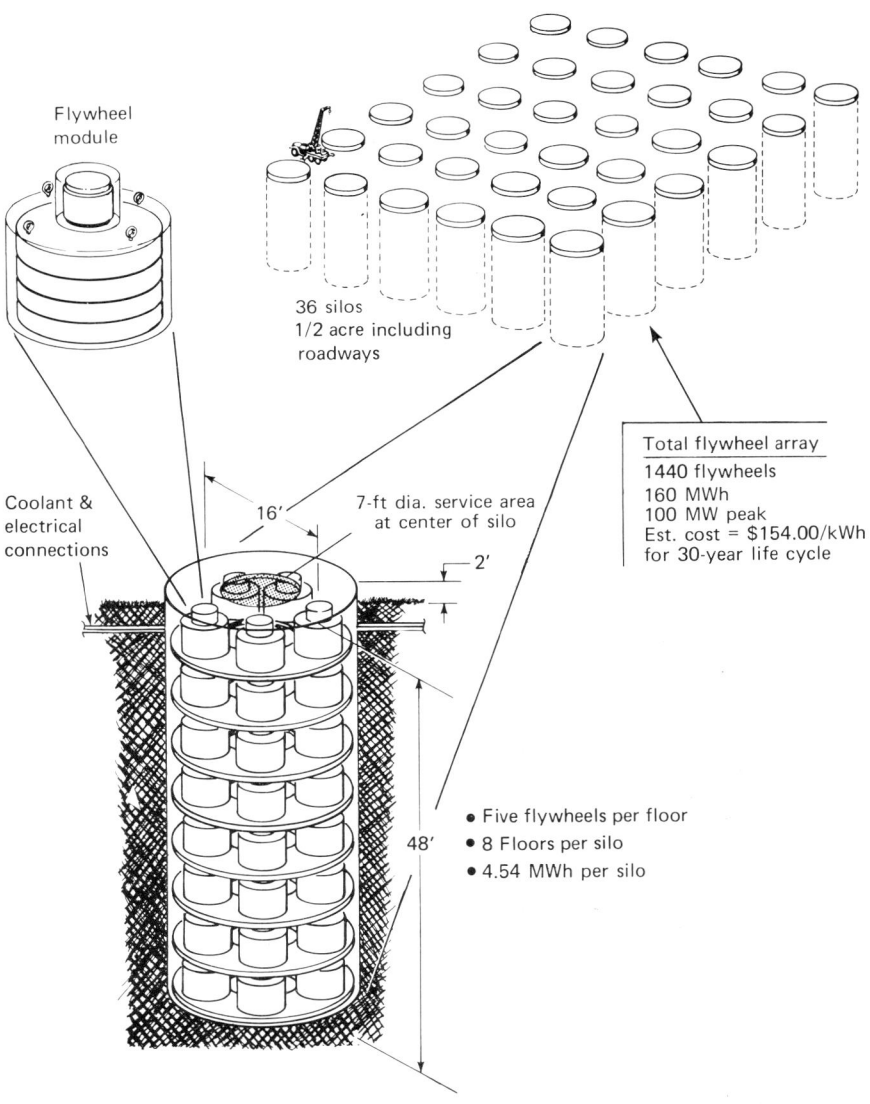

Flywheel
module

36 silos
1/2 acre including
roadways

Coolant &
electrical
connections

16'

7-ft dia. service area
at center of silo

2'

Total flywheel array
1440 flywheels
160 MWh
100 MW peak
Est. cost = $154.00/kWh
for 30-year life cycle

48'

• Five flywheels per floor
• 8 Floors per silo
• 4.54 MWh per silo

periods to set the flywheel spinning. At peak demand, the energy from the spinning flywheel could be used to generate electricity. A combined electric motor/generator is used to carry out both operations. The entire flywheel motor system would be arranged with a vertical axis in a special underground chamber in, say, a garage of the dwelling., A high vacuum would be maintained within the container housing the flywheel and bearings.

Flywheel energy storage also appears to have a future in road vehicles, particularly those involving frequent stop/start operation as in urban traffic. The basic idea is that when decelerating one does not convert valuable mechanical energy into heat energy by applying the brakes. Instead, the kinetic energy is stored by setting a flywheel spinning. The power surge needed for vehicle acceleration can be provided from the spinning flywheel, thereby obtaining optimum efficiency in the use of the energy. In petrol driven test vehicles operational economies of the order of 50 per cent have already been achieved in this way. It is expected that similar economies would be obtainable with electric vehicles.

Pumped water storage

Although in Great Britain the nuclear power capacity is only about 3–4 per cent, nuclear energy supplies about 15 per cent of the annual output of electricity. The reason for this is that nuclear power stations, whose main cost is the capital invested in them, are best operated at a very steady rate. Fuel costs with nuclear stations are a fraction of those incurred with fossil fuel fired electric generating stations. Since also it is not too easy to vary very drastically the output from nuclear power stations, it follows that it makes sense to use these stations whose fuel costs are low for generating the base power load.

Power stations fuelled by coal, gas and oil can be throttled back when loading falls, but even so it is best to steady the load if one can. There are also occasions when there are very high peak load requirements which may only last a few minutes a day. To satisfy these it would be necessary to have costly standby stations which would only operate for very short periods over the year. Yet not to satisfy them may mean voltage reductions and the accompanying danger that motors and other equipment may burn out due to overheating.

As the percentage of nuclear capacity in relation to the total installed electric generating capacity rises, these problems will be accentuated. A first attempt to solve them is the pumped storage scheme at Dinorwic in North Wales. There are two existing lakes near the town of Llanberis, the Llyn Peris lake which lies in the valley and the Marchlyn Mawr lake, which lies about 500 m above it on top of a mountain. Both these lakes have been enlarged and are connected to each other by a system of tunnels with enough capacity to drive six 313 MW turbine generators.

When these turbine generators are driven in the opposite direction, powered by electricity from the National grid they act as motor pumps. They then consume 281 MW of electricity per unit.

When in operation, the Dinorwic station will be able to provide a constant output of 1680 MW for five hours during peak demand periods, using the potential energy of the difference in water level between the upper and lower reservoir to drive the generators. It then takes the pumps six hours to transfer the water once more from the bottom Llyn Peris lake to the top Marchlyn Mawr lake. This is done at night using cheap off-peak electricity.

At periods of sudden surge of power demand the Dinorwic station can be put into operation far more rapidly than almost any alternative system. It is claimed that it can reach an output of 1320 MW within ten seconds from scratch. This means that the system can be put into operation extremely rapidly if there is a sudden voltage drop due to power consumption momentarily outstripping production.

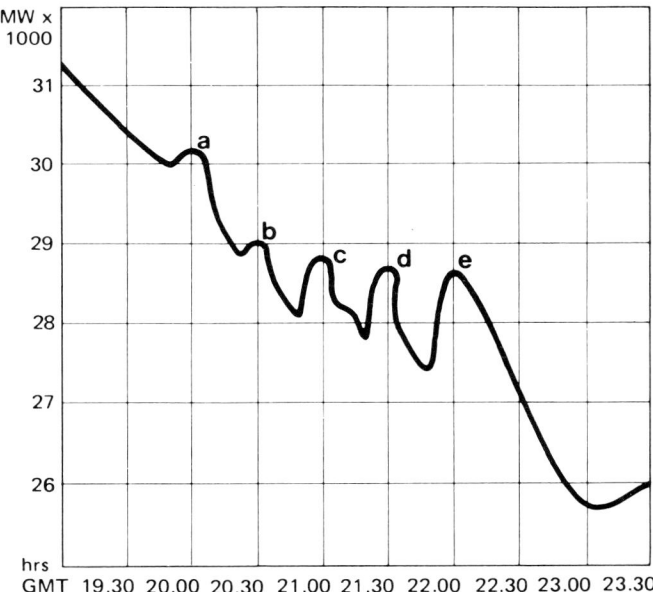

6.14 Variation of electricity demand. Courtesy Central Electricity Generating Board
The effect of a popular television programme on electricity demand is apparent in this curve, which shows demand peaks occurring during the evening of 28 October 1975 when the film 'Dr No' was shown on all ITV Regions. Peaks a and e coincide with the start and finish of the film; peaks b,c and d with commercial breaks.

Detailed description of scheme

The Dinorwic scheme is being designed and built by the Generation Development and Construction Division of the CEGB, of Barnwood, Gloucester, the consultants for the civil engineering work being James Williamson and Partners, Glasgow and Binnie and Partners, London.

During a full generating cycle Dinorwic will use more than 6.6 million m³ of water. This is to be retained in the enlarged Marchlyn Mawr reservoir by a rock fill dam 600 m long, which is landscaped on the downward face to blend with the scenery. The upstream side is faced with asphalt to provide the necessary water seal and the flexibility to meet the pressure changes caused by such a vast weight of water being continually moved in and out of the lake. The dam contains about 1.85 million cubic metres of rock fill and is about 35 m higher than the previous water level of the lake.

Underground works

Water from Marchlyn Mawr will flow through the hydraulic tunnels at a maximum rate of 390 m³/s. It is carried to the turbines via inlet tunnels three km long, including a vertical shaft 440 m long and 10 m diameter. To create the station's network of huge tunnels and caverns it was necessary to excavate no less than two million tonnes of slate. The underground chamber housing the main plant is one of the largest excavated caverns in the world, measuring 180 m × 23.5 m × 52 m.

There are six generator/pumps in the main plant cavern, two of which would be kept spinning in air at no load to provide an immediate reserve for the network. The machines have been designed to stand being loaded and unloaded up to 40 times per day. Electricity is generated at 18 000 V in the machine hall and is conducted by aluminium bus bars to the transformer hall, measuring 154 m × 23.5 m × 18 m, where the voltage is stepped up to 400 000 V. From there the power is fed into the national grid at that voltage.

After passing through the turbines the water flows out through three tail race tunnels which discharge below the surface of Wellington Pool, a flooded part of the slate quarry which is connected to the Llyn Peris lake. Four million m³ of slate and rock waste tips had to be removed to enlarge Llyn Peris from 37 hectares to 57 hectares to provide the necessary volume of water. The water level of Llyn Peris will only rise and fall by 14 m when the station is operating at maximum capacity so that it was only necessary to build embankments 3.64 m in height.

Dinorwic is the largest pumped power station in Europe and follows on from the much smaller station in nearby Ffestiniog, which has a maximum power storage capacity of 360 MW. The civil engineering works are designed for a life of 80 years and the engineering plant for 40 years. The entire scheme was begun early in 1974 and is expected to be fully operational shortly.

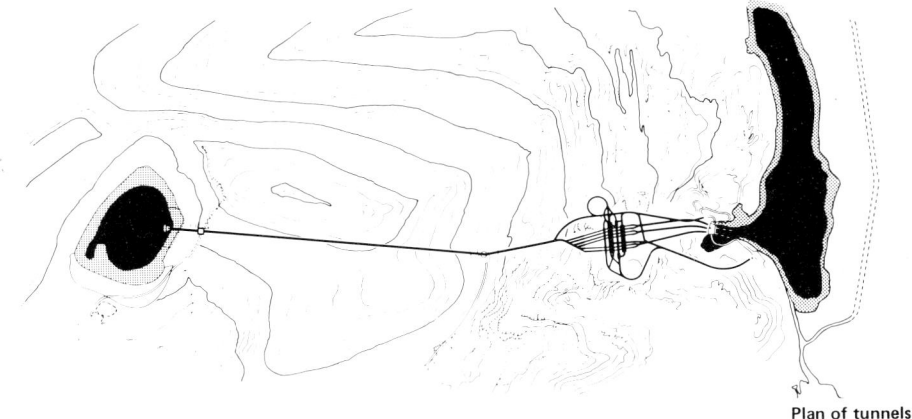

Plan of tunnels

Section

6.15 Plan and section of Dinorwic site

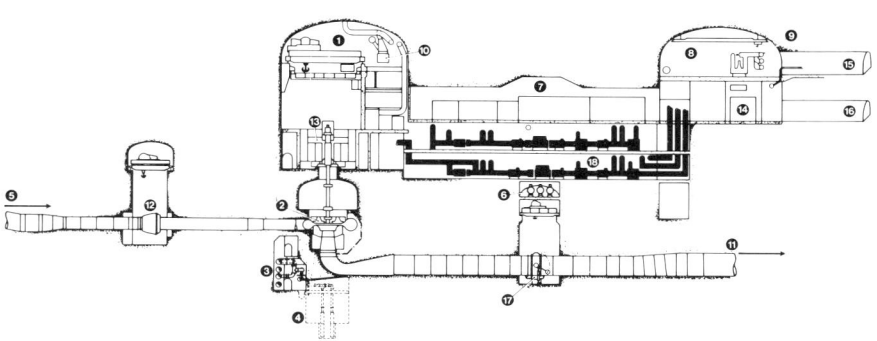

6.15 (continued) Section through Dinorwic power station.

1 Machine hall	10 Heating and ventilation plant
2 Pump turbine	11 To tailworks and lower reservoir
3 Pipe gallery	12 Main inlet valve
4 Drainage sump	13 Generator motor
5 From top reservoir	14 Generator motor transformer
6 Starting busbars	15 Cable tunnel
7 Busbar gallery	16 Plant access tunnel
8 Transformer hall	17 Draught tube valve
9 400 kV switchgear	18 Metal clad switchgear

Steam storage

The classic example of thermal energy storage in the form of steam is to be found in the Berlin Charlottenburg CHP power station. Sixteen vertical steam storage vessels with a diameter of 4.5 m and a height of 21 m are filled with superheated water. Steam is obtained from these vessels by flashing off when the pressure is released. These storage vessels were first established in 1929 but were found most useful in the post-war period, as West Berlin is an island economy and not connected to any power grid system. In case of turbine failure the steam stored in these tanks can restart a turbine without the need for any external power supply.

Work is being carried out in West Germany on the construction of prestressed cast iron vessels for energy storage in the form of superheated water used in conjunction with nuclear reactors. These vessels consist of cast iron blocks, lined on the inside with steel and thermally insulated on the outside. They are prestressed by both circumferential and axial high-tensile tendons. Often these vessels carry thermal insulation materials on the inside as well. The design is intended to produce storage chambers able to store hot pressurized water at temperatures of up to 300°C and pressures of 100 bars, avoiding the weaknesses engendered by welding. So far vessels with internal diameters of 7.6 m, internal heights of 45.5 m and volumes of 2000 m^3 have been built, which can store up to 75 MWh of electrical energy. It is considered that energy storage vessels of this type could probably be designed to store up to 300 MW h of steam energy. Such methods may well be a reasonably economic system of levelling power demand and supply, which will be an increasing problem once an appreciably larger fraction of electrical energy is produced by nuclear power than fossil fuel.

SELECTION CHART: ENERGY STORAGE

Type of storage	Advantages	Disadvantages	Appropriate applications
Sensible heat	Cheap materials	Limited capacity	Domestic purposes
Latent heat	High efficiency thermal regeneration	Higher cost	Domestic and small scale industrial purposes
Chemical	Excellent recovery	High cost	Industrial processes and large scale domestic systems
Compressed air	Mechanical energy storage suitable for power plants	Large scale only	Still experimental

SELECTION CHART: ENERGY STORAGE

Type of storage	Advantages	Disadvantages	Appropriate applications
Flywheels	Potential small-scale electrical energy storage	Still expensive	Eventually likely to be used for domestic purposes Still in pilot plant stage
Pumped water	Excellent large-scale electric power storage	Only suitable for very large projects	For public utility companies
Steam	Suitable power storage system for power stations	Limited applicability	For public utility companies

References and further reading

1 J. Silverman, *Energy storage*, Pergamon, Oxford, 1980.
2 S. W. Yuan and M. M. Majdi, 'The mini-prototype solar energy earth storage system', *Energy* 6, July 1981, pp 571–584.
3 C. Wyman, J. Castle and F. Keith, 'A review of collector and energy storage technology for intermediate temperature applications', *Solar energy* 24, 1980, pp 517–540.
4 P. D. Metz, Numerous papers on ground storage of thermal energy, Brookhaven National Laboratory, Upton, NY USA.
5 C. D. MacCracken, *Salt hydrate thermal energy storage system for space heating and air conditioning*, Calmac Manufacturing Corporation, Englewood, New Jersey USA.
6 Technical Information from Billings Corporation, Independence, Missouri, USA.
7 Technical information on Dinorwic scheme from Central Electricity Generating Board, North West Region, Europa House, Birdhall Lane, Cheadle Heath, Cheshire.
8 Proceedings of International Conference on Energy Storage, Brighton UK 29th April–1st May 1981 BHRA Fluid Engineering, Cranfield Beds. UK.
9 Z. Stunic, V. Djurickovic and Z. Stunic, Thermal storage nucleation of melts of inorganic salt hydrates, J. Appl. Chemical Biotechnology, 28, 1978, pp 761–764.
10 Numerous articles: International Journal of Hydrogen Energy.

SELECTED COMPANIES INVOLVED IN MANUFACTURE OF PRODUCTS

UK and Europe

Dornier System GmbH, Postfach 1360, 7990 Friedrichshafen 1, West Germany.

Siempelkamp Giesserei GmbH, Siempelkampstrasse 45, Postfach 45, D- 4150 Krefeld 1, West Germany.
Société Electrique de l'Our, Rue Pierre d'Aspelt 2, Luxembourg.

United States

AIDCO Maine Corp., Orrs Island, ME 04066.
Berry Solar Products Div., 2850 Woodbridge Avenue, Edison NJ 08837.
Calmac Manuf. Corp., 150 S. Van Brunt Street, PO Box 710, Englewood NJ 07631.
Caloskills Div. Girton Manuf Corp., Millville, PA 17846.
Megatherm, a Vapour Corp Div., 803 Taunton Avenue, East Providence, RI 02914.
LPC Inc., PO Box 37, New Richland, MN 56072.
OEM Products Inc., Airport Ind. Park, Plant City, FL 33566.
Sunmaster Corp., 35 W. William Street, PO Box 1077, Corning, NY 14830.
Thermal Energy Storage Inc., 10637 Roselle Street, San Diego, CA 92121.
Trisolar Corporation, 10 De Angelo Drive, Bedford MA 01730.

7 Heat recuperators

Heat recuperators, or heat exchangers, as they are also called, are pieces of equipment which can abstract sensible heat from one stream of flowing fluid and supply it to another stream. They are an essential feature of all production processes in the chemical industry. Because of the importance of improving heat recovery, consquent on the very appreciable rise in prime energy costs, heat exchangers are becoming increasingly important in the heating and ventilating field as well.

Principal current uses
1 To extract useful heat from waste hot liquids and gases. The heat is transferred to secondary fluids, which can then be used for either space heating or for the supply of hot water to kitchens and bathrooms.
2 To operate calorifiers, which are particularly widely used in the district heating field. Thermal energy is transferred from the circulating fluid, which has had to be dosed with posionous substances such as hydrazine, morpholine and caustic soda, in order to protect mild steel pipes from corrosion. The heat is given off via heat exchangers to highly purified town water to enable it to be used for cooking and washing purposes.
3 In district and group heating practice, heat exchangers are used to provide indirect hot water supply to, for example, high buildings. The supply hot water may be at a pressure insufficient to enable it to service either the top floors of a tall building or one sited on top of a hill. In such cases it is advantageous to use water/water heat exchangers to transfer the heat to the secondary medium, which can then be pumped to the top by a separate system.
4 For normal heat transfer from steam heaters or flues to circulating air, in order to raise this air to the required working temperature.
5 For the operation of air conditioning equipment, in which heat is being abstracted from room air by the refrigeration fluid or by chilled air.
6 For the supply of heating to swimming pools, where heat generated by either conventional heat sources or by solar batteries is transferred to the large volume of swimming pool water.
7 For heat recovery from exhaust air, flue gases and other sensible heat sources.

Classification of heat exchangers

Heat exchangers can be subdivided conveniently into three categories:
1 Liquid/liquid heat exchangers
2 Liquid/gas heat exchangers or gas/liquid heat exchangers.
3 Gas/gas heat exchangers

LIQUID/LIQUID HEAT EXCHANGERS

The most common system in use is one in which one fluid flows through a series of parallel pipes, while the other passes through the space between the tube bundle and the casing which contains this tube bundle. The heat transfer surface is the total circumferential area of all the tubes. The hot medium can flow either in the tube bundle or in the casing which surrounds it. The real criterion of choice is the liability of either of the two fluids to cause scaling or other deposition of impurities. Cleaning out the tube bundle (calandria) is a reasonably easy procedure for most heat exchangers of this type, and for this reason the dirtier fluid or the one more liable to scaling is always allowed to flow through the calandria. The cleaner fluid goes across the tube bundle.

An alternative design to this shell and calandria system is one in which there are a number of parallel plates, with alternate spaces being occupied by the fluid which gives up its heat, and the one which receives it. Heat exchange between the two fluids follows the basic equation 7.1 below:

$$dQ = dA\,U\,(t_h - t_c)\quad J/s(W) \tag{7.1}$$

where dQ is the heat transferred through an area of dA m^2. The U-value between the two flowing fluids, including the interface (metal plus encrustations), is U   W/m^{2}K, and the temperatures of the hot and cold flowing fluids are t_h and t_c °C respectively. The reason why one selects differential values for Q and A is because the values of U and $(t_h - t_c)$ vary at different positions of the heat exchanger. Obviously, when operating any heat exchanger we seek to maximise the value dQ. Equation 7.1 shows clearly that the heat transferred is a product of the following three variables:

1 *The area of interface between the two flowing liquids*
All heat exchangers should be designed in such a way that, within the limits of costs and overall sizes of equipment, the interface area is a maximum. For example, let us consider the case where there is a choice between a calandria consisting of 20 tubes with an internal diameter of 40 mm, or a calandria consisting of 80 tubes with an internal diameter of 20 mm. Both tube bundles will fit into the same size of external casing which has a length of 2.5 m.
In the former case the interface area is:
$20 \times \pi \times 0.04 \times 2.5 = 6.283$ m^2

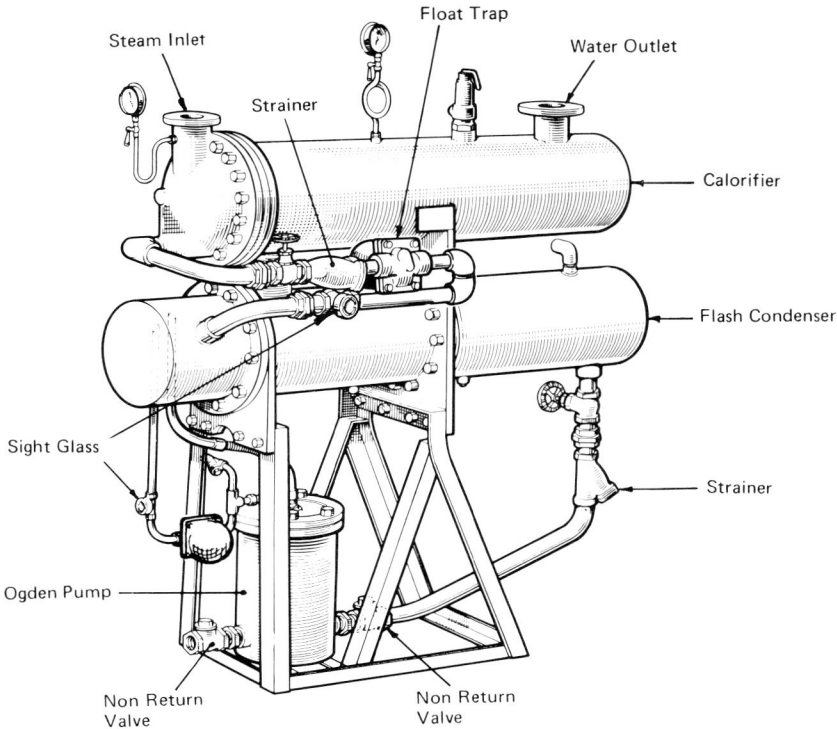

7.1 Packaged calorifier/flash condenser unit. Courtesy Spirax Sarco

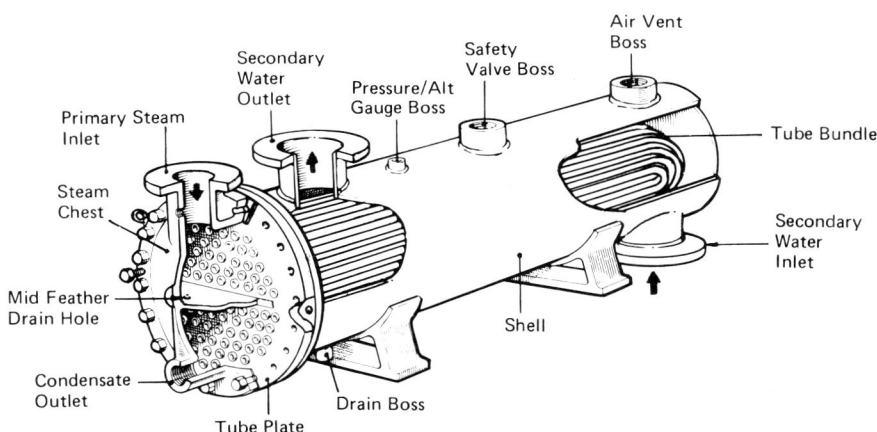

7.2 Non-storage heating calorifier. Courtesy Spirax Sarco

In the second case the interface area is:
$80 \times \pi \times 0.02 \times 2.5 = 12.567 \text{ m}^2$
Everything else being equal, the second heat exchanger should be twice as efficient as the first.

2 The U-value of the interface
This consists of three sections:
1 The hot side laminar layer
2 The solid interface
3 The cold side laminar layer
In the absence of scaling and other deposition of substances with a poor thermal conductivity, the thermal conductance of the metal interface is very high. The main sources of thermal resistance are the two laminar layers. If fluids pass past the interface surface with streamline motion (ie Reynolds number is below 2000) these laminar layers constitute appreciable thermal resistance. The faster the flow of fluid past the interface surfaces, the thinner the stagnant laminar layer is going to be, and in consequence the better the heat transfer. Improvements in reducing the laminar layer are achieved by the introduction of baffles and other devices to increase turbulence.

3 The temperature difference between the two fluids
Heat transfer between the hot fluid and the cold fluid of a heat exchanger is aided by an appreciable temperature difference between the fluids. It must not be forgotten, however, that a high temperature difference constitutes a rapid write-off of high-grade energy and should therefore be avoided.

Co-current and counter-current heat exchangers
It is possible to operate heat exchangers in two ways, co-current and counter-current. In the co-current mode the hottest part of the heat-donating fluid give off its heat to the coldest part of the heat receiving fluid, so that at the beginning of the heat exchange process $(t_h - t_c)$ is a maximum. As the heat exchange process continues, so the value of $(t_h - t_c)$ is reduced, being a minimum at the outlet from the heat exchanger. Such a system is best suited to small and cheap heat exchangers as there is a rapid heat exchange at the initial position, which drops off rapidly as the $(t_h - t_c)$ values falls. It is obvious that co-current heat exchangers are not economical to use if a very high efficiency of heat exchange is needed. They are also unsuitable for long flow paths, as the cost of a longer heat exchange surface becomes unviable.

Co-current heat exchangers should be squat in design, with the maximum heat exchanger area at the position where the fluids enter. At this position too, one should seek to obtain a maximum U-value by the induction of adequate turbulence. Co-current heat exchangers are most suitable under the following circumstances:
a When it is desired that as much heat as possible should be transferred from one fluid to the other, using equipment which is as cheap as possible to buy and maintain, but where it is not vital that all the heat or a very high proportion of it is used.

b Where high viscosity of the fluids makes it impractical to induce high turbulence over a considerable length of the heat exchanger, because this would raise pumping costs to intolerable levels.

In the counter-current mode of operation the hottest inflow faces the warmest outflow. This means that throughout the heat exchange operation, the value of ($t_h - t_c$) remains essentially constant. By and large the efficiency of such heat exchangers is directly proportional to their length and the surface area of the calandria.

Heat exchangers operating on the counter-current principle are much easier to design than those on the co-current principle and are mainly used for the following purposes:

a When it is necessary to transfer as much heat as possible from the heat donating fluid to the heat receiving fluid, using a costly and complex heat exchanger plant.

b Where the difference in temperature between the fluids is low.

c When the temperature of the heat donating fluid leaving the heat exchanger is lower than the temperature of the heat receiving fluid leaving the heat exchanger.

For example, it is quite feasible with counter-current heat exchangers to have a heat donating fluid entering the heat exchanger at, say, 150°C and leaving the exchanger at 80°C, while the heat receiving fluid is heated up from 40°C to 120°C or even more. This would, of course, be impossible to achieve with co-current operation.

Laminar and turbulent flow
In streamline flow, liquid molecules flow along in a parallel fashion and in consequence heat transfer from the centre of the fluid to the walls of

TABLE 7.1 THERMAL CONDUCTIVITIES OF VARIOUS SOLIDS AND LIQUIDS

Material	Thermal conductivity W/m K at 20°C
Aluminium	237
Copper	166
Iron	147
Magnesium	159
Silver	427
Zinc	115
Water	1.964
Toluene	0.44
Petrol	0.47
Glycerol	0.97
Oil	0.75
Air (no convection)	0.025

the heat exchanger tube proceeds by conduction only. As **table 7.1** shows, thermal conductivities of fluids are remarkably poor compared with those of metals.

It is therefore always necessary to ensure that fluids in heat exchangers move turbulently: ie in such a fashion that constant mixing occurs.

When turbulent motion occurs, one can accept that the entire body of the fluid has the same temperature because of the turbulence. The only conduction heat transfer which is needed is across the boundary layer. This is a layer of stagnant fluid between the turbulent fluid and the wall of the tube or other interface which contains it. Turbulence can be induced in a fluid if Reynolds' number exceeds about 2000. Reynold's number is commonly quoted as:

$$N_{Re} = \frac{v \, d \, D}{\mu}$$

where v is fluid velocity in m/s
      d is diameter of pipe containing it or other linear dimensions for different systems in m
      D is density of fluid in kg/m^3
      μ is viscosity of fluid in kg/m s

It can be seen that if the various units are entered as given, Reynolds' number is a dimensionless function: it is the same when imperial units are used as it is when the newer SI units are employed.

Example: Determine the minimum speed necessary for water at 80°C to attain turbulent flow in pipes of internal diameter 15 mm.

Let us assume that the density of water at 80°C is 971.83 kg/m^3 and that its viscosity is 3.45×10^{-4} kg/m s.

If $v_{(min)}$ is the minimum velocity required for turbulent motion to occur we can write that:

$$2000 = \frac{v_{(min)} \times 0.015 \times 971.83}{3.45 \times 10^{-4}}$$

or $v_{(min)} = 0.0473$ m/s

What this means in practical terms is, that if individual tubes are of 15 mm diameter, then the water speed must be kept at more than about 50 mm/s for the achievement of viable operation as a heat exchanger. If the water speed falls below this value, the heat exchanger becomes virtually ineffective.

The more turbulent the fluid flow, the thinner the laminar layer becomes and in consequence the better the heat transfer between fluid and wall. The mathematics of the actual relationship between Reynolds' number and the heat transfer is somewhat involved. An approximation can be given in that the thickness of the laminar layer, once turbulent conditions of flow have been achieved, is roughly proportional to the inverse of the Reynolds' number, taken to a power of 0.8. What this means in practice is, that provided other conditions such as temperature of fluid, viscosity and density are kept constant, the heat transferred between the body of the fluid and the wall varies with $v^{0.8}$.

Theoretically therefore, the higher the speed of the fluid past the wall, the better the heat transfer between the fluid and the wall will be. Equally, the receiver fluid should also move very turbulently and at a high speed relative to the wall, to facilitate easy take-up of heat.

The problem is, however, that high fluid speeds require a large expenditure of pumping energy, cause erosion and excessive corrosion inside the heat exchangers and thereby raise capital expenditure. As in every other field of engineering design, it is therefore necessary to compromise.

One seeks to achieve the maximum degree of turbulence, and thus the minimum thickness of stagnant fluid layers which provide the main resistance to heat transfer, at a minimum expenditure on pumps and pumping.

Increased turbulence can sometimes be achieved at a comparatively low cost of increased pump head by incorporating baffles, or inducing eddies by spiral structures or other means.

To sum up, the skill of the heat exchanger designer is to be found in his ability to produce increased turbulence and therefore thinner fluid laminar layers at a minimum cost in lost pumping head.

Calculation of heat transfer area needed
The area of heat transfer surface depends upon the overall heat transfer coefficient between the primary and secondary fluid, the average temperature difference between these fluids and the amount of heat which needs to be transferred. Obviously, the most economical heat exchanger designs are those in which one can accommodate as much heat exchanger surface as possible within the smallest volume. However, this may set problems from the cleaning and pumping point of view and one has to make a compromise.

The basic equation for heat transfer is given in equation 7.2:
$$Q = U A (\Delta T) \tag{7.2}$$
where Q is the heat transferred per unit time, expressed usually in watts (J/s), U is the overall heat transfer coefficient in W/m^2 K and ΔT is expressed in kelvins, while A, the area of heat transfer surface is given in m^2.

Calculation: determine the size of heat exchanger needed if the primary flow water amounts to 100 l/h with inlet temperature of 130°C and outflow temperature of 70°C. This serves to heat up the secondary flow of water from 20°C to 80°C. It is assumed that counter-current heat exchange is employed. The specific heat of water varies with the temperature but is asssumed to amount to 4.186 J/g K over the range of 70–130°C. Find the area of heat exchanger surface needed and the volume flow of the secondary water, assuming that the average U-value of the heat exchanger interface equals 466 W/m² K.

Solution: the average temperature of the primary heating fluids equals:

$$\frac{(130 + 70)}{2} = 100°C,$$

while the average temperature of the secondary fluid flow equals:

$$\frac{(20 + 80)}{2} = 50°C.$$

This gives the average temperature difference between the primary and the secondary flow as:

$$100 - 50 = 50°C.$$

As the primary heating water is being cooled down from 130°C to 70°C (through a temperature range of 60°C), the secondary water is heated up through the same range; 80°C − 20°C = 60°C. Primary and secondary water flows are the same ie 100 litres per hour.

 If the secondary water flow is larger than the primary one, as for example, in swimming pool heaters, then the temperature rise achieved is going to be correspondingly smaller.

The following equation always applies:

$$v_{pr} (\Delta T_{pr}) = v_{sec} (\Delta T_{sec}) \tag{7.3}$$

where v_{pr} and v_{sec} are the primary and secondary water flow rates in litres per hour, while (ΔT_{pr}) and (ΔT_{sec}) are the corresponding temperature differences in K.

 The heat which is being supplied from the primary fluid to the secondary fluid equals: (less extraneous heat losses):

$$100 \times 4.186 \times 50 \times 1000 \text{ J/h}$$

$$= 2.093 \times 10^7 \text{ J/h} = \frac{2.093 \times 10^7}{3600} \text{ J/s} = 5813.9 \text{ W}$$

The heat exchanger area needed is therefore:

$$\frac{5813.9}{466} = 12.48 \text{ m}^2.$$

 To minimise heat losses to the outside air, it is best to design heat exchangers in such a way that they are rather squat. Designs of this type have the minimum external surface area per unit volume or fluid throughout.

It has already been pointed out that counter-current heat exchangers must be rather long to enable good heat exchange to take place. To reconcile the above two requirements one often uses a twin- or multi-pass system of heat exchange. In this the liquid in the tubes passes backwards and forwards against the flow of the other liquid. Multi-pass systems produce some problems with regard to the cleaning out of the calandria, but can be used satisfactorily when the more scaling fluid is used on the outside, or when chemical cleaning is found to be sufficient. The shells of all heat exchangers have to be extremely well insulated, otherwise heat losses can make operation uneconomic.

Typical commercial water/water heat exchangers

Spiral heat exchangers

Spiral heat exchangers such as the ones made by the Israeli firm of Lordan and Co. and marketed in the UK by D. J. Neil Ltd. of Macclesfield, Cheshire, have very much higher heat transfer coefficients than equivalent straight tubes. An empirical equation gives the heat transfer coefficient U_{sp} to be related to the U-value for equivalent straight pipes U_{st} in the following way

$$\frac{U_{sp}}{U_{st}} = 1 + (3.5 \ d/D) \tag{7.4}$$

where d is the tube diameter in mm while D is the spiral diameter in mm.

The Lordan spiral tubes usually have a d/D ratio of 0.263 so that the heat transfer coefficient is virtually twice as high as with equivalent straight tubes. The units have the additional advantage that due to their very squat shape, and the fact that they are always of single-pass design, they need not be arranged horizontally. Equivalent straight tube exchangers have to be positioned in such a way to avoid gas bubbles accumulating. Gas bubbles are always scrubbed through, even when the heat exchangers are arranged vertically. The fluid with the higher scaling factor must always be placed on the shell side.

Mechanical cleansing is difficult, but a technique of squeezing the soft calandria bundles to dislodge deposits, combined with chemical cleaning, has been developed. Spiral heat exchangers are used in the heating and ventilating industry for heating domestic hot water by means of either hot district heating water or steam.

The Lordan heat exchanger scores particularly when a very compact unit is needed due to lack of space.

The Swedish AGA-CTC AB is one of the most important companies in the field, supplying a whole range of heat exchangers for the district heating industry. Their M range consist of straight U-tube calandria with a two-pass system having a maximum working pressure of either 6 or 13 bars.

These units are made in a variety of materials including mild steel, copper, brass, copper-lined steel and stainless steel. For the district heating industry such heat exchangers are mainly employed for abstracting heat from the main network to supply a closed system intended to provide heating to a tall block of flats or a high-lying district. Heating surfaces of up to 90 m² are made with 250 mm supply lines and take-off lines. The mild steel shell is surrounded by thick layers of glass wool mats for insulation, covered by lacquered fairing plating.

Another range of units made by the firm employs spirally assembled helical tubes with, in some cases, cross-spiral indentations. These indentations have two purposes:

1 To increase turbulence and thus the heat transfer coefficient.

2 To increase the strength of the tube against external pressures.

The design of these heat exchangers, which are used in the main as calorifiers to supply domestic hot water from circulating district heating water, permits the use of fluid temperatures of up to 150°C and pressures of up to 16 bars. The heat exchangers are insulated on the outside with 50 mm thick mineral wool slabs and are finished with varnished timber slats, which give them an attractive appearance. U-values

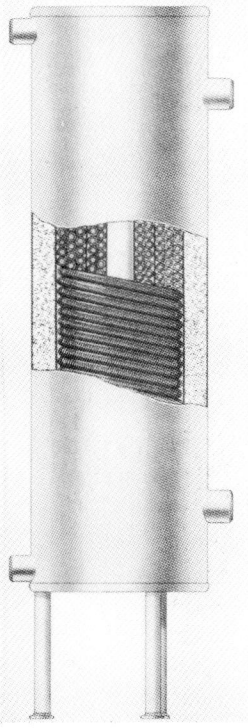

7.3 The CTC SKR heat exchanger

claimed for water/water heat exchangers are 466 W/m^2 K and for steam/water heat exchangers 930 W/m^2 K.

Tacotherm Ltd construct special water/water heat exchanges for heating up swimming pools. In this case the primary heat flow water is heated by coal, oil or gas fired boilers. This flows through the jacket of a twin-pass smooth pipe heat exchanger system consisting of only one single U-tube. The boiler water flows through the shell, while the swimming pool water is circulated through the U-tube. The unit is made in various sizes depending upon the amount of heat to be transferred.

The U-tube is of copper while the shell is made from seamless steel piping with cast iron connecting flanges. Maximum operating pressure is 10 bars with a maximum operating temperature of 190°C. The unit is resistant to chlorinated water.

Burner Products Ltd market very simple hot water cylinder heat exchangers. To counteract the fact that there is only natural convection on the secondary side, the surface area there is increased considerably by finning the outside surface of the U-tube heat exchanger, using a complex copper wiring system.

The primary hot water is pumped through the U-tube and the heat dissipated to the secondary fluid via the finned surface, which is

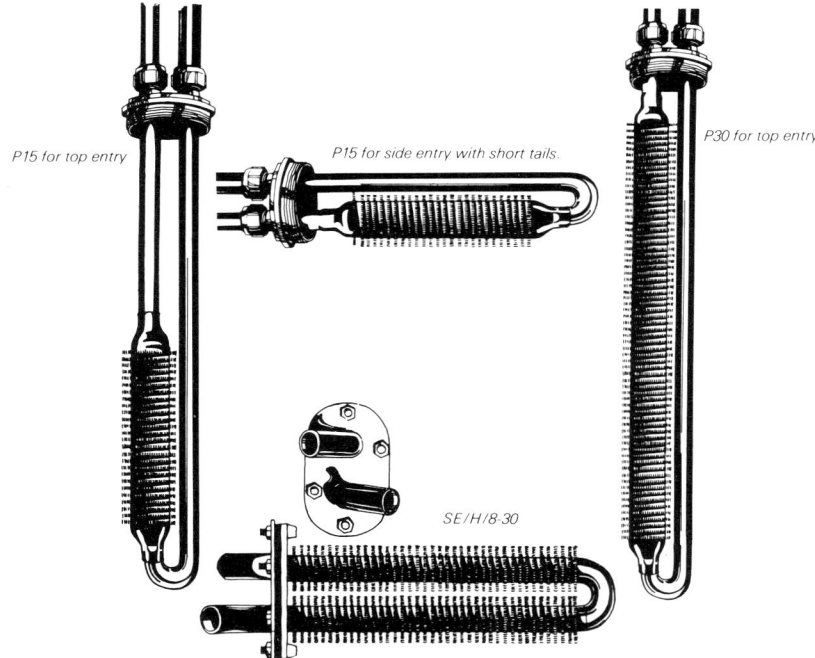

7.4 Burner Products' Econozone heat exchangers

attached to either one or both limbs of the U-tube. The trade name for the device is Econozone and the heat tube is 15 mm diameter copper tubing to be sealed by a fibre washer against the 60 mm boss supplied with the unit. The system is suitable for cylinders varying in height between 600 and 1200 mm.

The unit should be fitted as far down as possible in the cylinder to improve convection heat transfer in the secondary fluid. The internal surface of the copper tube base has a helical coil bonded to it, to increase the primary conductive heating surface and to make the primary water turbulent. The external surface comprises a copper wire winding secured to the base of the copper tube by strong wire binding. The whole is fused together with special capillary solder.

This presents a very large heat transfer surface to the water in the tank. This system of finning with an open wire winding has a heat conduction surface both parallel and at right angles to the water flow inside the tube. It is superior to more conventional solid fins.

Assuming that the temperature of the primary hot water is about 80°C while that of the water tank is to be maintained at an average of 65°C, between 2.3 and 8.8 kW are supplied to a hot water tank, depending upon the flow speed of the primary fluid.

The system can be used with solar panels, using water heated by solar energy to heat up the hot water tank, thereby reducing the loading on the electric immersion heater during the summer months. Econozone heat exchangers can also be used for other purposes such as transfering solar heat to water in swimming pools.

If the capacity of a single Econozone heat exchanger is insufficient, the manufacturers recommend the fixing of two or more such heat exchangers in parallel. Up to eight units can be fitted in this way. If such a system is connected to a single 28 mm manifold to give a supply of 13.6 l/min of water at 80°C, then a heat output of 73.3 kW can be obtained. Tests were carried out in hard water zones and also in areas where aggressive water conditions occur. It was found that provided tank water temperatures were kept below 65°C the wire element would keep clean for up to 7 years. The constant expansion and contraction of the wires in the element shakes off any hardness scaling formed, to produce a slurry which can be removed easily. A newer development is to nickel plate the units for use with acid and other aggressive waters.

Crosse Engineering Ltd manufacture a large number of types of heat exchangers. Most are of the steam/water type, as illustrated.

Water to water calorifiers made by the company employ either straight tubes or coils. They are made in sizes between 12.2 and 293 kW output with a standard duty of heating water from 10°C to 65°C with a primary mean temperature of 76°C.

Swimming pool calorifiers are based upon a primary hot water temperature of 81°C with a 71°C return, to maintain a swimming pool

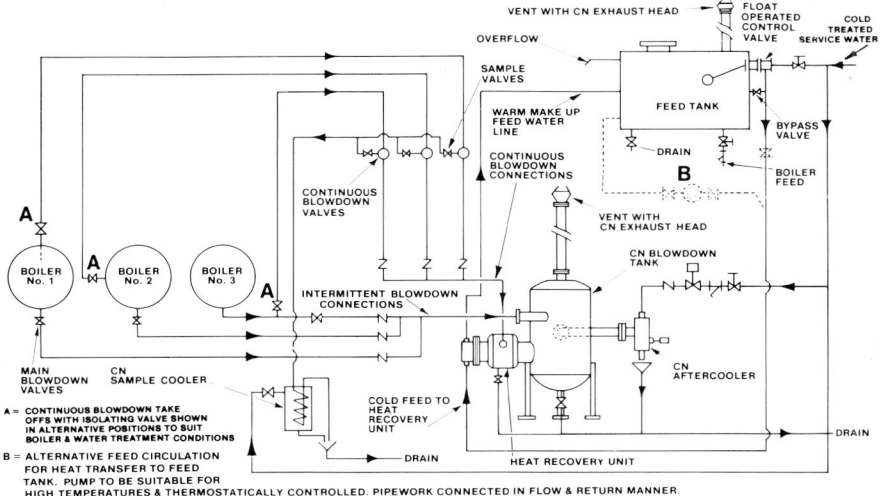

7.5 Curwen and Newbery continuous blowdown heat recovery unit

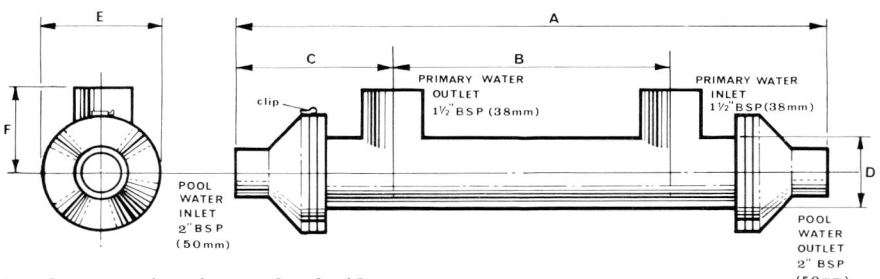

7.6 Crosse swimming pool calorifier

temperature of 22°C. The models made by the firm have ratings between 30 kW and 78 kW with a maximum pool water flow of 8800 l/h. For normal hard water the shell of the heat exchanger through which the primary water flows is made of mild steel, while the tube plate and tubes are of bronze-faced mild steel. For soft water the tubes are of rolled copper, while for salt water they are of aluminium brass.

Plate heat exchangers

Plate heat exchangers are mainly employed when both fluids tend to scale or foul their surfaces. It is a good deal easier to clean such units than the more traditional tubular systems because it is easy to dismantle plate type heat exchangers completely. Plate heat exchangers consist of standardized plates with a herringbone pattern stamped in.

AGA-CTC heat exchangers employ five different plate sizes which have surface areas varying between 0.08 m^2 and 1.7 m^2. Each standard plate is manufactured in two different thermal length modifications.

This thermal length is obtained by changing the pattern angle of the herringbone. Some of the heat exchange plates also employ a wash-board pattern. The plates are arranged in any flow configuration required to suit the particular application. The most common is, however, the pure parallel connection. It is even possible to use one plate heat exchanger for more then two fluids. A pasteurizing system as illustrated involves three different fluids: the product, the heating fluid and the cooling fluid. There are occasions when one plate heat exchanger can operate with as many as five different fluids.

Materials used for construction
The standardized plates can be made in any material that can be press-shaped. The most common materials are: stainless steel, titanium, Incoloy, Inconel, Hastelloy B and C and Carpenter 20 alloy. Titanium is always used for seawater applications while Hastelloy is used for fluids which contain free sulphuric acid.

Gaskets are used to seal the gaps between the plates as they are assembled in the frame. The most common material is nitrile, which is suitable for water and aqueous solutions up to a temperature of around 140°C. For steam and aqueous media above 140°C ethyl propane rubber (EPDM) is used. For more severe conditions gaskets are made from either viton or teflon. The heat exchanger plates are assembled in a frame of high carbon steel, using clamping bolts.

Plate heat exchanges have the following advantages and disadvantages as against more traditional tubular systems:

Advantages
Easily dismantled for cleaning and inspection.
Occupies minimum space.
Heat losses are low.
Heat transfer is very uniform and can be adjusted accurately with great flexibility.
It is easy to modify the system to increase or reduce heat transfer surfaces, without having to scrap existing equipment.

Disadvantages
Since gaskets must be used, pressures are limited to around 20 bars.
It is not advisable to use the system at temperatures above 170°C.
Costs are a good deal higher than with traditional systems. The system has a much shorter effective life than traditional tubular systems because of the need for gaskets.

Dimensioning or plate heat exchangers
Two values have to be calculated:
1 The NTU_{he} value which is dimensionless and describes the thermal length from the flow pattern point of view.

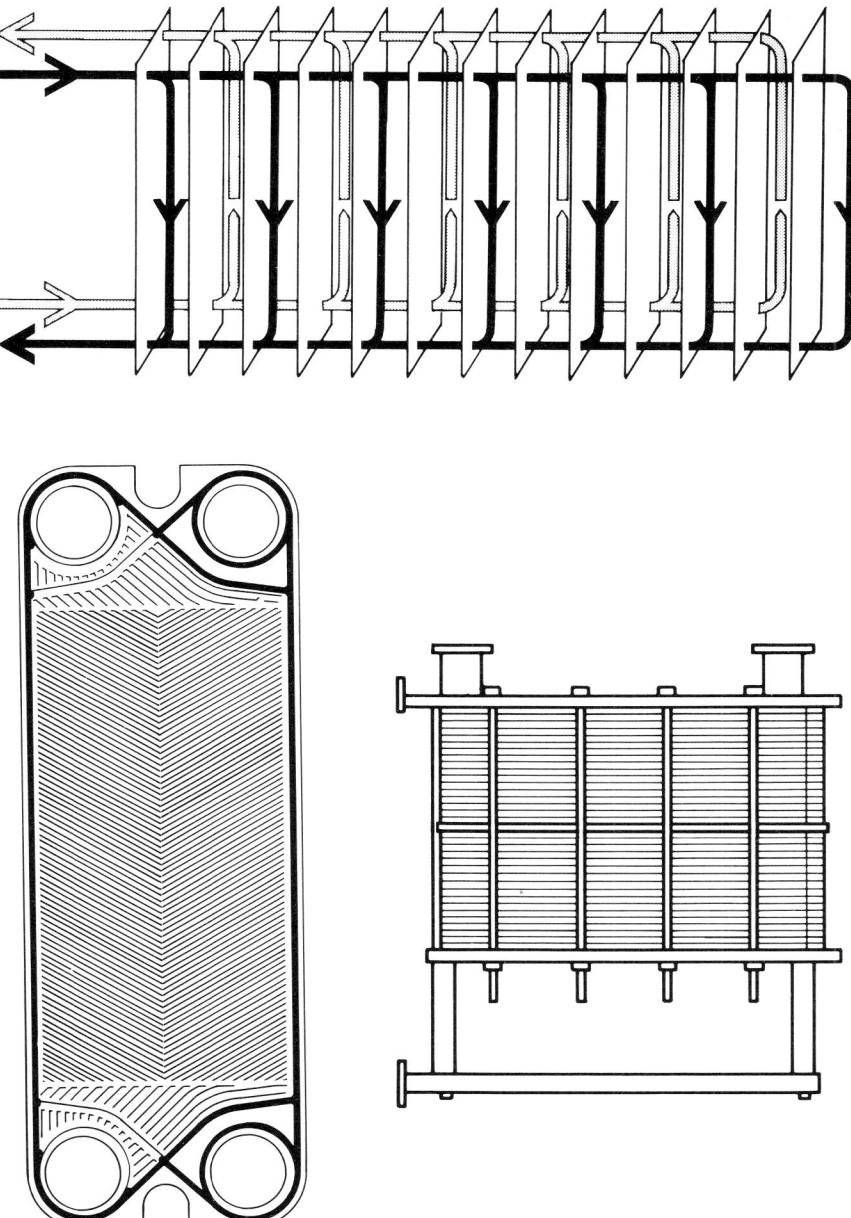

7.7 AGA-CTC liquid/liquid plate heat exchanger

2 The $NTU_{process}$ value which is also dimensionless and describes the thermal length in terms of the heat transferred.

The two equations (7.5 and 7.6) are:

$$NTU_{he} = \frac{U \, f}{D \, C_p \, v}$$ (7.5)

$$\text{and } NTU_{process} = \frac{T_1 - T_2}{ATD}$$ (7.6)

where: U   is U-value between media (W/m^2 K)
      f   is factor (d1) (depending on design)
      D   is density of medium (kg/m^3)
      C_p is specific heat of medium (J/kg K)
      v   is velocity of medium (m/s)
      T_1 is hot inlet temperature (K)
      T_2 is hot outlet temperature (K)
      ATD is average temperature difference (K)

This ATD value can be calculated as follows (7.7)

$$ATD = \frac{(T_1 - T_4) - (T_2 - T_3)}{\log \dfrac{(T_1 - T_4)}{(T_2 - T_3)}}$$ (7.7)

where T_1 and T_2 are the hot side inlet and outlet temperatures, respectively. T_3 and T_4 are the cold side inlet and outlet temperatures, respectively. For ideal dimensioning NTU_{he} should be as nearly as possible equal to $NTU_{process}$.

One of the main variables is the pressure drop between the inlet and the outlet of each flowing medium. The velocity of flow is directly proportional to the square root of this pressure drop. This means that the velocity of flow can be doubled by increasing the pressure difference fourfold. This, however, halves the NTU_{he} value. As fouling of surfaces increases, the U-value drops and in consequence NTU_{he} falls also. Once NTU_{he} has dropped to an uneconomic level, the plant must be stripped down and cleaned.

Actual calculations involving designs of plate heat exchangers are rather involved and require the use of specialist computer programs which have been developed by the various companies which market such devices.

Spiraflow heat exchangers

This system was invented in Australia a few years ago and is now marketed in Europe by AGA-CTC. The system consists of between 3 and 9 tubes arranged concentrically inside each other. The tubes are kept in position by two end castings in the form of stepped pyramids. Each tube fits on to its corresponding step.

The end casting also distributes the media to the annular passages. The heat exchanger is kept together by a central bolt and can therefore

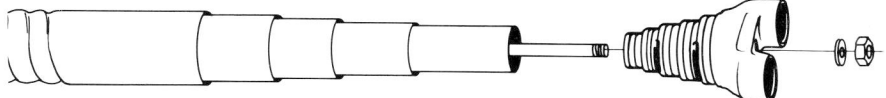

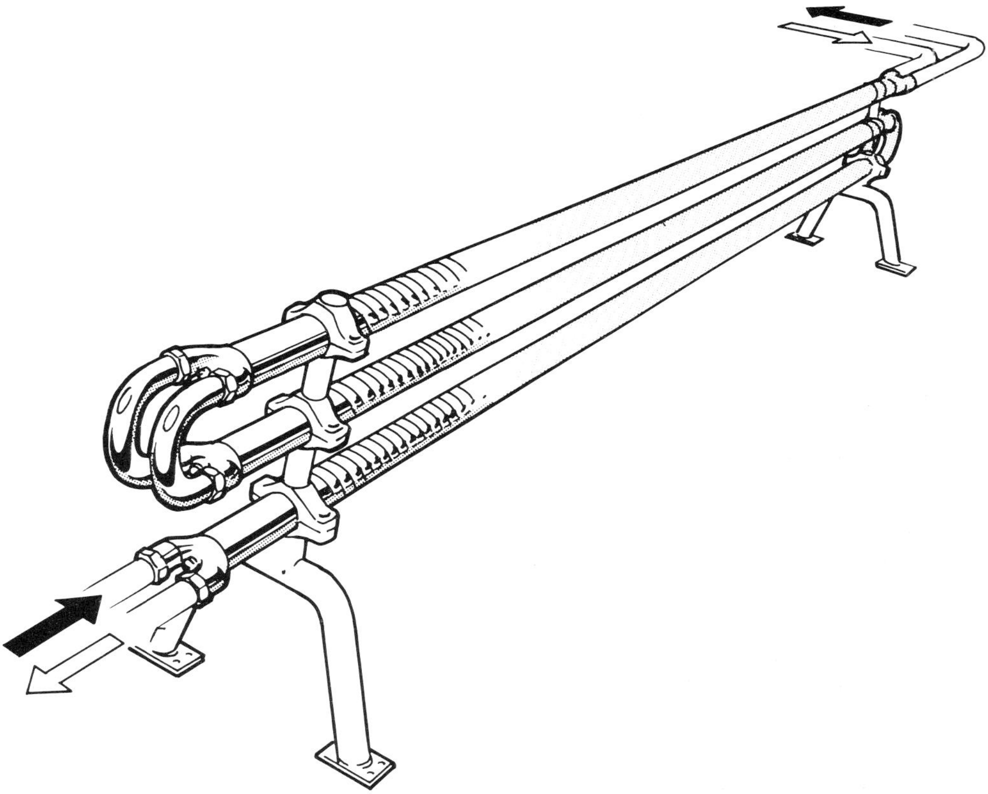

7.8 AGA-CTC Spiraflow heat exchanger

be dismantled with ease. Two O-rings against the tubes seal the annular passages against each other. The tubes, which are spirally corrugated to achieve better flow turbulence, vary in length between 1 m and 5 m. Both they and the end castings are always constructed from stainless steel. The innermost tube has a diameter of 50 mm, and each succeeding tube has a diameter 12.6 mm larger, so that the gap between tubes is 6.3 mm. The O-rings are made from either nitrile, EPDM, viton or teflon, depending upon the purpose for which the unit is to be used.

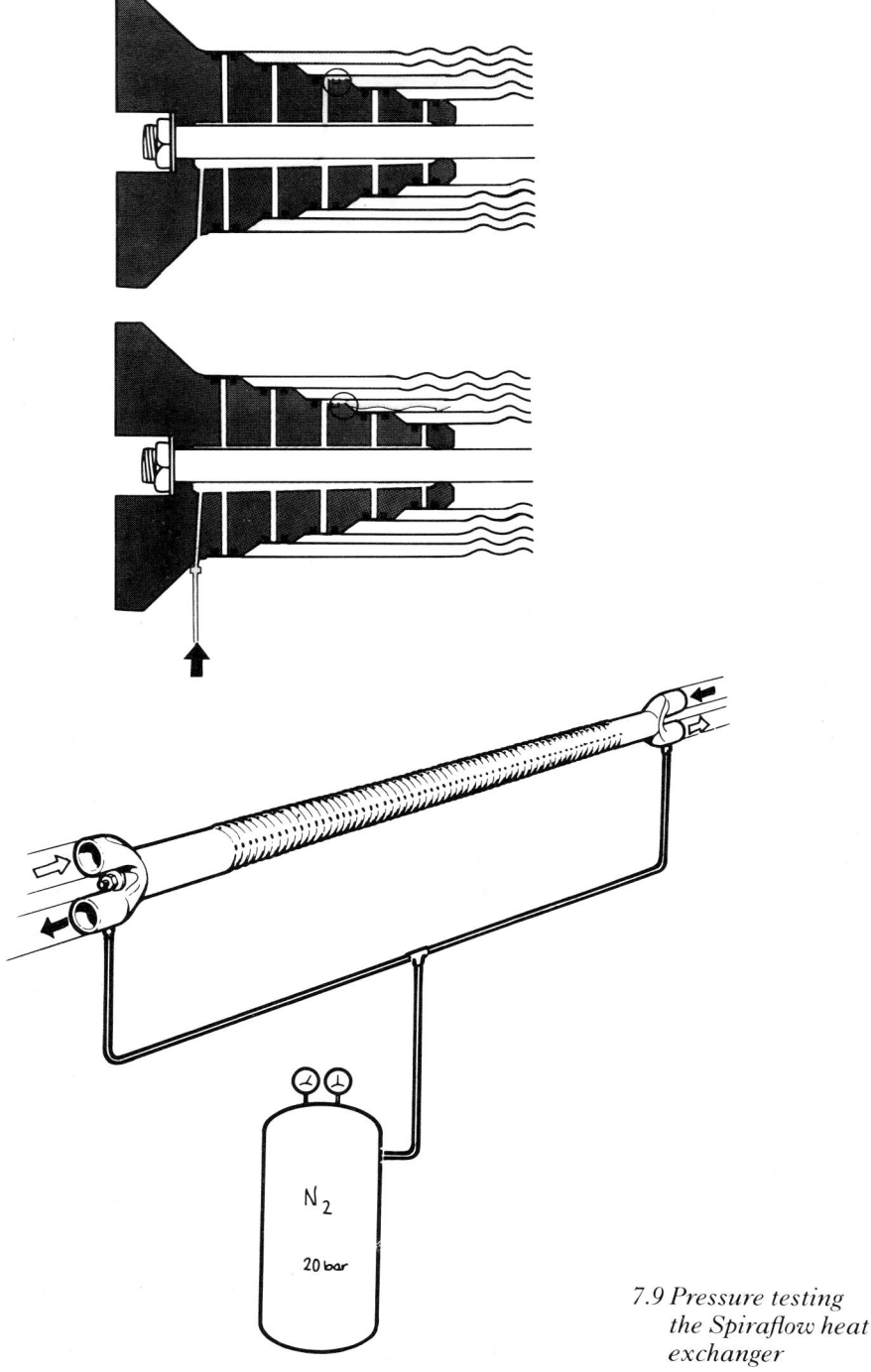

7.9 Pressure testing
the Spiraflow heat
exchanger

Advantages claimed

The unit is easy to dismantle and to clean, simply by releasing the central bolt.

The heat transfer rate is even and there are no dead pockets as with some of the more traditional designs. The unit can withstand pressures of up to 40 bars and temperatures up to 250°C. Special devices are incorporated to warn of any mixing between the two media. An inert gas is fed into the inner tube of the system at a higher pressure than the medium. If there should be a leak the pressure drop is monitored immediately and an alarm sounds.

Spiraflow heat exchangers can be used for the following purposes:
a Heat recovery from fluids which are either dirty or highly viscous and from liquids which contain a high percentage of solid particles or fibres.
b Heat exchange with fluids which are dangerous when mixed with the heat receiving fluid. This would include nuclear heat applications and fluids which may be bacterially infected.
c Heat exchange with slurries.

LIQUID/GAS AND GAS/LIQUID HEAT EXCHANGERS

We have already dealt with the flow of liquids in heat exchangers. Gases can be either non-condensible, such as air or, condensible, such as steam.

Non-condensible gases have a far poorer heat transfer than equivalent fluids and, in addition, a very small specific heat per unit volume. This means therefore that air and similar gases have to be blown at a high rate through heat exchangers with a very large surface area, to be at all effective. On the other hand, there is never any problem with regard to streamline motion at low speeds of flow, as would be the case with liquids. Heat exchangers in which the primary flow is a liquid such as hot water and the secondary flow is air include air heaters, air conditioning equipment and heat reclamation plants.

Heat exchangers in which the primary flow is a non-condensible gas and the secondary heat medium is a liquid, would be the economizers, in which hot flue gases heat up water for turbine feeds or for other purposes. Here again we find heat reclamation units. In these the sensible heat of exhaust air is removed by a fluid, and the heat re-supplied to incoming cold air.

Condensible gases like steam are far more effective as heat exchange media because of the considerable heat evolved when the gas changes into a liquid.

Condensible steam is a very important heating medium and is used in steam heaters. It is always necessary to provide means of run-off and drainage. The film heat transfer coefficient for vertically heated steam calandria is as high as $10\,000$ W/m^{2}K with some designs, so that

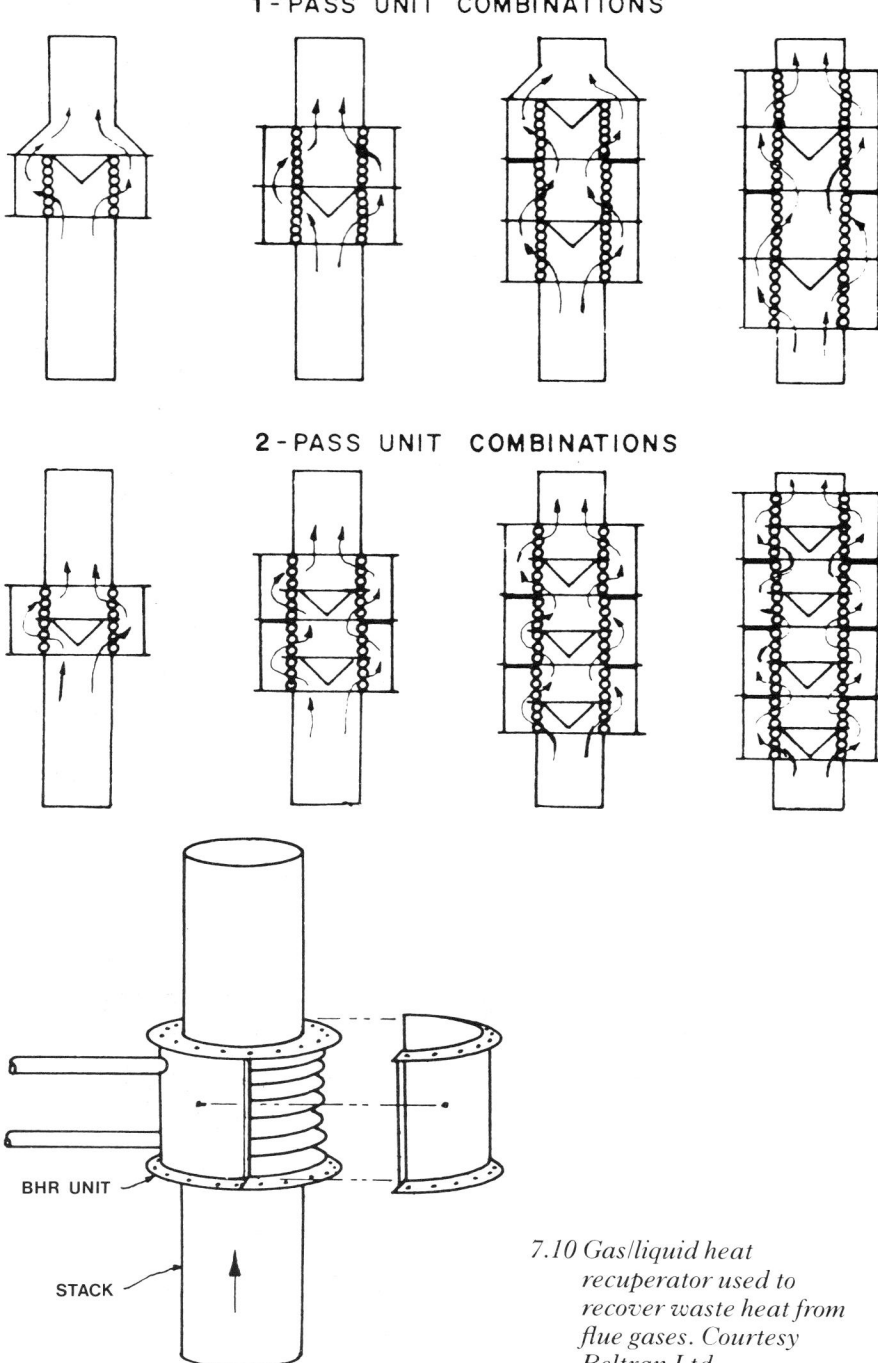

7.10 Gas/liquid heat recuperator used to recover waste heat from flue gases. Courtesy Beltran Ltd

extremely compact heat exchangers can be built. In fact, in such exchangers nearly all the resistance to heat exchange is on the tube/water side. This means, in effect, that a steam-heated heat exchanger of the same capacity as its equivalent water heated exchanger need only have half the interface area of the latter, if we assume that all other conditions such as temperature difference and secondary water speed are kept the same.

Uses of liquid/gas and gas/liquid heat exchangers

These types of heat exchangers are used for the following main purposes:

a For blowing hot air into environmental areas using hot water or other fluids as the source of heat.

b For operation of an air conditioning system in which air is cooled by chilled circulating water.

c For the extraction of sensible heat from exhaust air for heat recovery.

The main problem involved with liquid/air heat exchangers is the very poor thermal conductivity of air when stagnant. It is only 0.02423 W/m K at 0°C rising linearly to 0.03185 W/m K at 100°C. The corresponding figures for *stagnant* water are 0.567 W/m K and 0.672 W/m K, respectively. This means that heat exchangers which employ air as a heating medium must have very much larger contact surfaces than plain liquid/liquid heat exchangers. In addition, every effort must be made to reduce the thickness of the boundary layer of stagnant air, which adheres to the surface of the heat exchanger and constitutes the dominant resistance to heat flow. In other words, one must induce maximum convection currents.

7.11 The Economair gas/liquid heat exchanger. Courtesy Curwen and Newbery Ltd

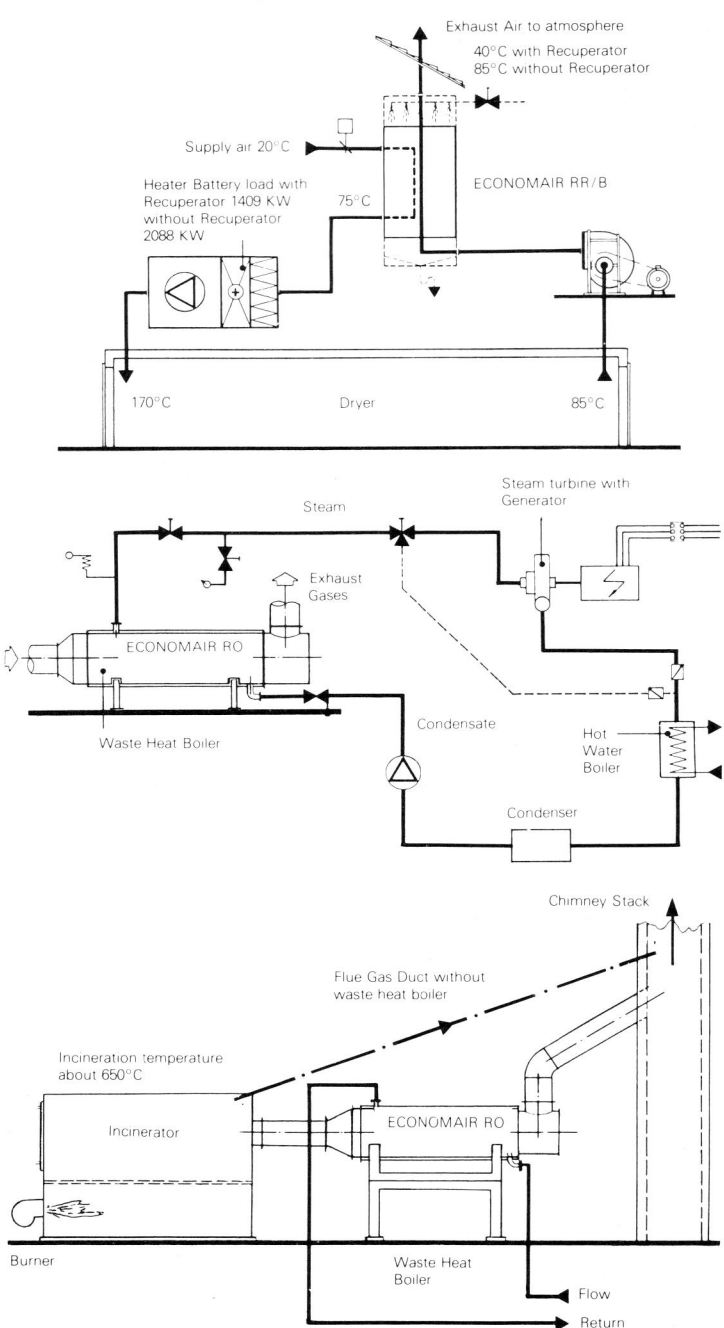

7.12 Industrial uses for Economair gas/liquid heat exchangers. Courtesy Curwen and Newbery Ltd

Free convection and radiation heat transfer

If there is any heated body surrounded by air, heat transfer from it takes place by both radiation and conduction/convection. Radiation generally passes straight through air and does not heat it, but transfers the heat directly to the next surface it lands on. Heat is supplied to the surrounding air from this surface which thus becomes a secondary convector. This is the basis of the ceiling-mounted black body radiant heating units which are now becoming increasingly popular due to the marked fuel economy which they achieve. The actual amount of heat transferred depends basically on the following parameters:

a The temperatures of the sending and receiving surfaces.

b The nature of the sending and receiving surfaces.

c The geometrical disposition of the sending and receiving surfaces.

The heat transferred from a plane surface 1 m^2 in area to another equal 1 m^2 surface arranged parallel to it, assuming that the receiving surface is a black body (ie is capable of absorbing all the heat it receives), while the emissivity of the surface sending out heat is e (d1) is given by the Stefan Boltzmann equation (7.8):

$$Q_r = 5.673 \times 10^{-8} \text{ e } (T_s^4 - T_r^4) \text{ watts} \qquad (7.8)$$

where Q_r is the radiant heat transfer in watts.

T_s is the surface temperature of the sending-out area in kelvins (°C + 273.15).

T_r is the surface temperature of the receiving area, also in kelvins.

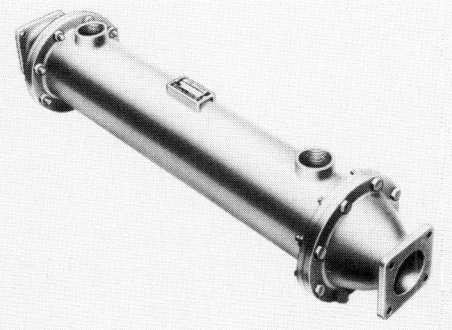

7.13 (left) Heating coil. Courtesy S&P Coils

7.14 (above) Bowman exhaust gas heat exchanger

Example: Let us assume that we have 1 m² heated radiator surface at a temperature of 100°C with an e value of 0.65, radiating its heat to an equal m² of a matt black surface (black body), kept at 15°C. Determine heat transfer by radiation only in watts:

$Q_r = 5.673 \times 10^{-8} \times 0.65 \ (373.15^4 - 288.15^4) = 460.7$ W

For other arrangements, such as a pipe surrounded by the walls of a room, or even more complex systems, various factors have to be derived and the basic Stefan Boltzmann equation multiplied by them. Equally, the equation becomes more complex when the receiving surface consists of a material which is not a black body or, worse still, consists of sections of various materials. Experimental measurement of radiation output is usually the only answer in such cases.

Free convection

Free convection currents occur when air is heated at a surface and in consequence its density is altered, causing colder and therefore denser air to replace it. The most generally useful equation used for such a purpose is the Heilmann equation (7.9):

$$Q_c = \frac{C \ (t_s - t_o)^{1.266}}{d^{0.2} \left(\dfrac{t_s + t_o}{2} \right)^{0.181}} \ \text{W/m}^2 \tag{7.9}$$

where Q_c is equal to heat transferred by free convection only.

t_s is the surface temperature in °C.
t_o is the external temperature in °C.
d   is a diameter function in metres, used for pipelines and other cylindrical bodies. For flat surfaces or in cases where d exceeds 610 mm, one writes d = 0.61.

C varies for different conditions:

Horizontal cylinders   C = 2.909
Long vertical cylinders   C = 3.536
Vertical plates   C = 3.992
Hot horizontal plates facing upwards or cold horizontal plates facing downwards   C = 5.125
Hot horizontal plates facing downwards or cold horizontal plates facing upwards   C = 2.548

Example: Determine the heat transferred per m² surface area by free convection only from horizontal pipelines of diameter 150 mm, if the surface temperature is equal to 80°C while the temperature of the surrounding air is 20°C.

Solution: Using the Heilmann equation we get:

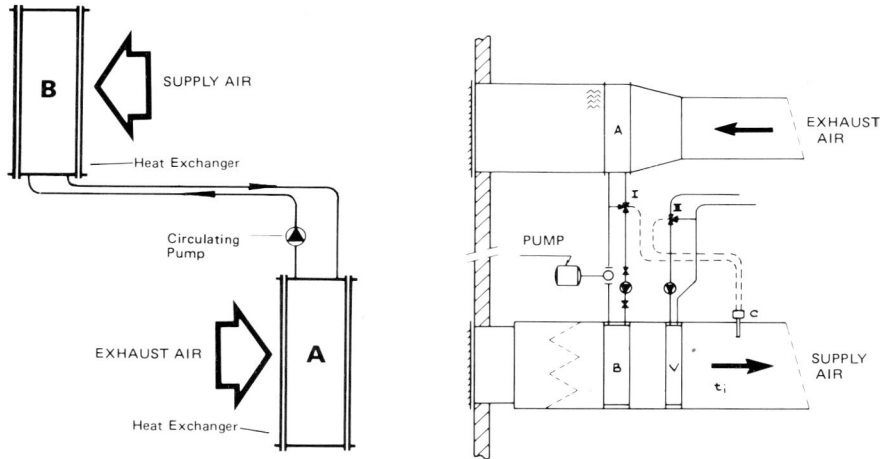

7.17 Basic concept of twin coil run-around heat recovery system. Courtesy Curwen and Newbery Ltd

if fitted. With run-around coils one can absorb heat from flue gases at the top of a chimney or from the exhausted vapours of a drying furnace high up in a building. This sensible heat can then be transported to a quite different position, which may even be in a different building, where sensible heat is required for preheating incoming air or for other purposes.

GAS/GAS HEAT EXCHANGERS

All gases including air have a very low specific heat and are also poor thermal conductors. In consequence air/air heat exchangers require very considerable interface areas and appreciable turbulence to restrict the thickness of the dead laminar layer. Virtually the only type of air/air heat exchanger used commercially is a multiple plate system.

In this system the supply air and exhaust air are completely separated from each other, being pumped at considerable speed past thin metal plates.

Air/air heat exchangers compete with thermal wheels when applied to the recovery of heat from exhaust air streams. There are two main fields in which they are superior to thermal wheels:

1 When handling very damp high temperature exhausts to warm up incoming dry air. Thermal wheels would get waterlogged under such circumstances and in consequence soon become quite useless. In the case of air/air heat exchangers, there is always provision for disposal of condensed water. If condensation takes place in an air/air heat exchanger the efficiency actually increases, because heat transfer is improved due to the utilization of latent heat released as water vapour in the air condenser.

2 When handling exhausts which contain extremely obnoxious gases, solvents and odours. Although thermal wheels have purge sections, these may not be fully effective.

The interface leaves from which the air/air heat exchangers are constructed are usually of aluminium. This is satisfactory for most designs

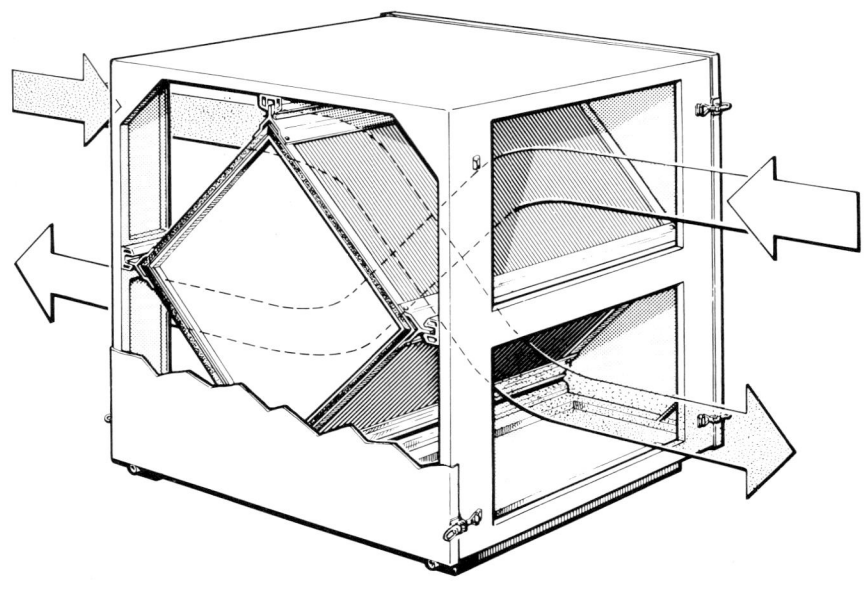

7.18 (above) Econovent cross-flow gas/gas heat exchanger

7.19 (left) A Curwen and Newbery heat recuperator

except those where the exhaust air is particularly corrosive. Under such circumstances the surfaces in contact with the exhaust gases are coated with a plastic film. In all air/air heat exchangers provision is made for easy dismantling to facilitate cleaning.

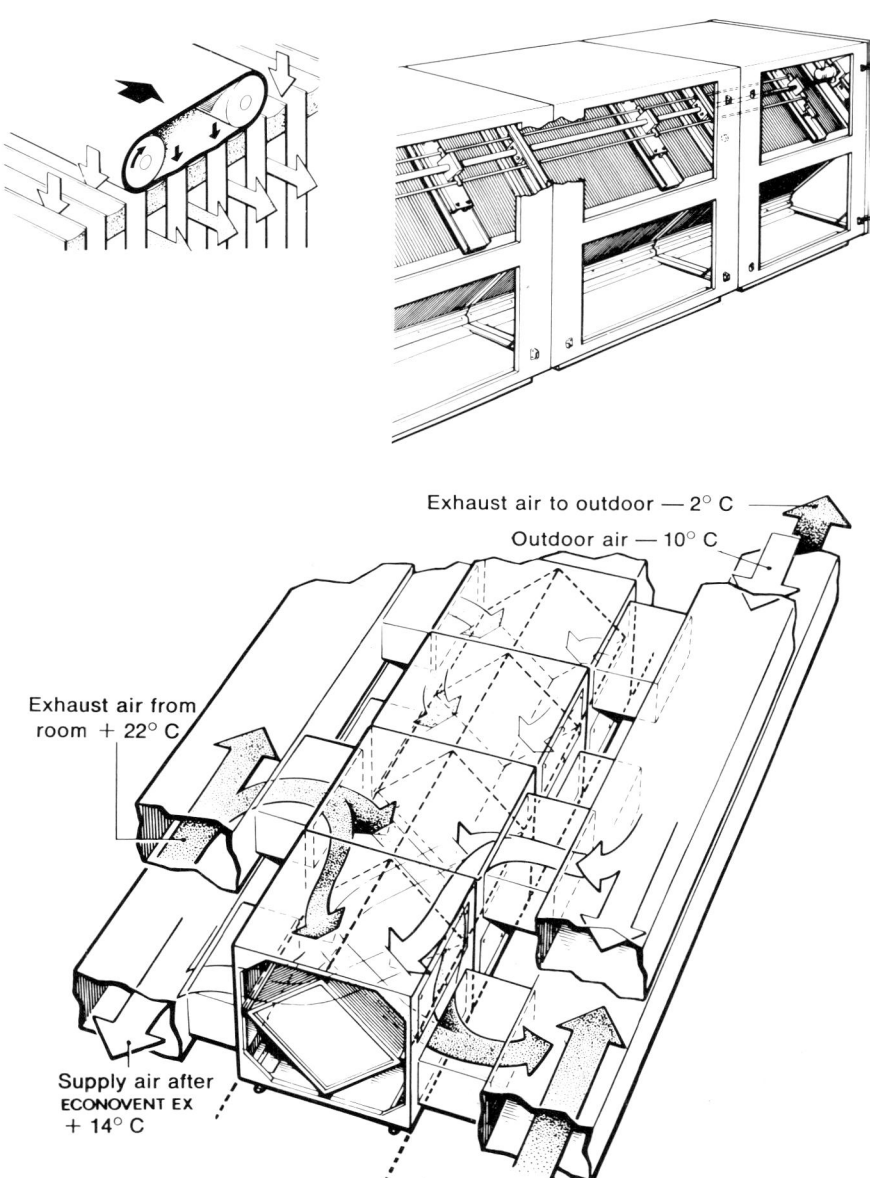

7.20 *Design details of the Econovent*

SELECTION CHART: HEAT RECUPERATORS

Type of equipment	Advantages	Disadvantages	Appropriate applications
Liquid/liquid co-current heat exchangers	Rapid heat transfer Low first cost Low pumping head	Poor efficiency of transfer Thermodynamically poor	Highly viscous fluids Where only partial heat recovery is needed
Liquid/liquid counter-current heat exchangers	Good efficiency of heat transfer Incoming secondary temperature may be higher than out-going primary temperature	High pumping costs Expensive to purchase	In cases (eg hot water calorifiers) where efficient heat transfer from one fluid to another is needed
Liquid/gas heat exchangers	Transfers heat from circulating liquid to air or other gases	Poor efficiency of heat transfer on gas side	Run-around coil systems
Gas/liquid heat exchangers	Excellent heat transfer Suitable for use with dirty gases	Sometimes liable to corrosion	Removal of heat from flue gases and other waste gases
Gas/gas plate heat exchangers	No mixing of dirty and clean gases possible Good heat exchange	Competes with heat regenerators More costly per unit output than regenerators	Heat recovery from exhaust air in cases where even slight gas mixing would be disastrous (hospitals)

References and further reading

1 W. H. McAdams, *Heat transmission*, McGraw-Hill, New York, 1954.
2 W. M. Rohsenow, *Developments in heat transfer*, E. Arnold, London, 1964.
3 H. Schenk, *Heat transfer engineering*, Longmans, London, 1960.
4 J. R. Simonson, *An introduction to engineering heat transfer*, McGraw-Hill, New York, 1967.
5 Technical literature issued by the firms mentioned in the text.

SELECTED COMPANIES INVOLVED IN MANUFACTURE OF PRODUCTS

UK and Europe

ACR Heat Transfer Ltd, Rollesby Road, Hardwick Industrial Estate, King's Lynn, Norfolk PE30 4LN.
Advance Services Ltd, Unit 2, Hamm Moor Lane, Weybridge Trading Estate, Weybridge, Surrey.

Auchard Development Co. Ltd, Old Road, Southam, Leamington Spa, Warks CV33 0HP.

Ballofix Ltd, Bishopsgate Works, Union Road, Oldbury, West Midlands B69 3ES.

Behncke GmbH, 8011 Putzbrunn, Werner von Braun Strasse 1, Germany.

F. H. Biddle Ltd, Newtown Road, Nuneaton, Warks CV11 4HP.

Bitzer Kuhlmaschinenbau, Eschenbrunnstrasse 15, D-7032 Sindelfingen, West Germany.

E. J. Bowman Ltd, Chester Street, Birmingham B6 4AP.

Burner Products Ltd, 74 London Road, Kingston-upon-Thames, Surrey KT2 6PZ.

Climate Equipment Ltd, Newby Road, Hazel Grove, Stockport SK7 5DA.

Clipper Air Handling Units Ltd, Raans Road, Amersham, Bucks HP6 6HY.

Conservatherm Ltd, Hereford House, Station Road, Billinghurst, West Sussex RH14 9SE.

Crosse Engineering Ltd, Herriott House, North Place, Cheltenham, Glos GL50 4DS.

CTC Heat Exchangers, 37 Lowlands Road, Harrow, Mddx HA1 3AW.

Delchi SpA, Via R. Sanzio 9, 20058 Villasanta, Italy.

Delta Ltd, Hollands Road, Haverhill, Suffolk CB9 8PT.

Exhausto A/S, DK-5550, Langeskov, Denmark.

Gledhill Water Storage Ltd, Sycamore Trading Estate, Squires Gate Lane, Blackpool, Lancs FY4 3RL.

Gunter Warmeaustauscher GmbH, Postfach 120, 8034 Germering, West Germany.

Hall-Thermotank Products Ltd, Chiltern Trading Estate, Grovebury Road, Leighton Buzzard, Beds LU7 8TU.

U. Hausmann & Söhne, 5100 Aachen-Brand, Birkenstrasse 2, Germany.

Haynes Coils Ltd, Telford Way Industrial Estate, Kettering, Northants.

Heat Transfer Ltd, Hadley House, Bayshill Road. Cheltenham, Gloucestershire GL50 3SP.

Kiloheat Ltd, Vestry Estate, Sevenoaks, Kent TN14 5EL.

Lindsey Engineering Group Ltd, Unit D, Riverside Industrial Estate, Riverside Way, Dartford, Kent.

Iain Miller Associates, Effie Road, London SW6 1EL.

Oy Nokia AB, PO Box 44, 01511 Vantaa 51, Finland.

Recuperator SpA, 20020 Lainate MI, Via Mantova 8, Italy.

S&P Coil Ltd, Evington Valley Road, Leicester LE5 5LU.

tt Coil APS, Svanningeved 2, 9220 Aalborg Øst, Denmark.

Tacotherm Ltd, Morison House, Rankine Road, Daneshill Estate, Basingstoke, Hants, RG24 0PH.

Wallsend Slipway Engineers Ltd, Point Pleasant, Wallsend, Tyne-and-Wear NE28 6QN.

Westwarm Ltd, Unit 6, Hither Green, Clevedon, Avon BS21 6XT.

Willison Controls Ltd, Dallas Road, Bedford MK42 9ES.

United States

Addchek Coils Inc., State Road, 66, Rte. 2, Forte Mill, SC 29715.

Aerco International Inc., 159 Paris Avenue, Northvale, NJ 07647.

Air Control Industries Inc., 213 McLemore Street, Nashville, TN 37203.

Air Dynamics Inc., 1918 N. Potrero Avenue, South El Monte, CA 91733.

Aldrich Co., East Williams Street, Wyoming, IL 61491.

American Standard Heat Transfer Division, 175 Standard Parkway, Buffalo NY 14227.

Ametek Heat Transfer Division, 2300 W. Marshall Drive, Grand Prairie, TX 75051.

Basco Division American Precision Industries Inc., 2777 Walden Avenue, Buffalo, NY 14225.

Berner International Corporation, R.D. 3 Wilmington Road, PO Box 5205, New Castle, PA 16105.

Blissfield Manufacturing Co., 626 Depot Street, Blissfield, MI 49228.

Chase Industries Inc., 8100 Reading Road, Cincinnati, OH 45222.

Coil Co. Inc., 125 S. Front Street, Colwyn, PA 19023.

Coil, Tool and Die Co., Highway 69 North, PO Box 1509, Jacksonville, TX 75766.

Dean Products Inc., 985 Dean Street, Brooklyn, NY 11238.

Emax Inc., 720 Grandview Avenue, Columbus, OH 43215.

Heatron Inc, PO Box 54, York, PA 17405.

HXK Inc., PO Box 27, Haworth, NJ 07641.

Kathabar Systems Ross Air Systems Division, PO Box 791, New Brunswick, NJ 08903.

Keep Rite Inc., 44 Elgin Street, PO Box 460, Brantford, Ontario N3T 5P4, Canada.

Koldwave Div. of Heat Exchangers Inc., 3100 N. Monticello Avenue, Skokie, IL 60076.

Kutrieb Corporation, 430 Phillip Street, PO Box 4303, Chetek, WI 54728.

Marcraft Inc. Div of Brasch Manuf. Co. Inc., 99 Ford Lane, Hazelwood, MO 63042.

Marley Cooling Tower Co., 5800 Foxridge Drive, Mission, KS 66202.

Nortec Solar Industries Inc., PO Box 698, Ogdensburg, NY 13669

Pyradyne Inc., 300 Nichols, PO Box 190, Hutchins, TX 75141.

Sing-Air Inc., 400 W. Walnut Street, PO Box 424, Gardena, CA 90247.

Standard Refrigeration Co., 2050 Ruby Street, Melrose Park, IL 60160.

Sun-Econ Inc., R.D 3 Barney Road, PO Box 426, Clifton Park, NY 12065.

Super Radiator Div. McQuay Group, 6714 Walker Street, Minneapolis, MN 55426.

Thrush/Amtrol Inc., W. 8th at Jefferson PO Box 228, Peru, IN 46970.

Turbotec Products Inc., 533 John Downey Drive, New Britain, CT 06501.

United Air Specialists Inc., 4440 Creek Road, Cincinnati, OH 45242.

Wedj Inc., 160 S. Hartman Street, PO Box 3485, York, PA 17402.

Young Radiator Co., 2825 Four Mile Road, Racine, WI 53404.

8 Heat regenerators

Waste heat from exhaust gases can be recovered by using either heat recuperators or heat regenerators. The former term encompasses the whole range of heat exchangers and the like, while the term heat regenerators implies the presence of an intermediate phase.

Basically what happens in a heat regenerator is that exhaust heat is absorbed by a solid thermal storage material. This heat is then given off to the incoming fresh air supply. The classical method of using heat regeneration is used in the gas industry to make hydrogen, hot producer gas is passed through chequer brickwork. This is raised to incandescence by the exothermic reaction which takes place when carbon monoxide burns to carbon dioxide. After a certain time the flow of producer gas and air is switched to another chequer brick chamber, and a mix of steam and carbon monoxide is passed through the incandescent chamber. The following reaction, which is endothermic (absorbs heat) takes place:

$$CO + H_2O \rightarrow H_2 + CO_2$$

The heat required to complete the reaction is taken from the incandescent bricks. The reaction proceeds until the brickwork has virtually cooled down. The flow of gases is then switched once more, with the exothermic reaction:

$$CO + \tfrac{1}{2}O_2 \rightarrow CO_2$$

heating up the cooled chequer brick chamber, while the heated chamber enables the endothermic reaction to take place. Many other industries also make use of such regenerative processes.

Today heat regeneration also plays an important part in simple energy conservation practice. Waste heat is taken off exhaust gas streams. It is then transferred to clean and fresh external air, thus reducing the load on the heating system and saving an appreciable amount of precious fuel.

THE THERMAL WHEEL

The thermal wheel heat regenerator was invented and patented originally by the Swedish engineer Frederick Ljungstrom in 1922. The first of these pre-heaters were made in Sweden, but as far back as 1924 James Howden and Co made some models in Scotland.

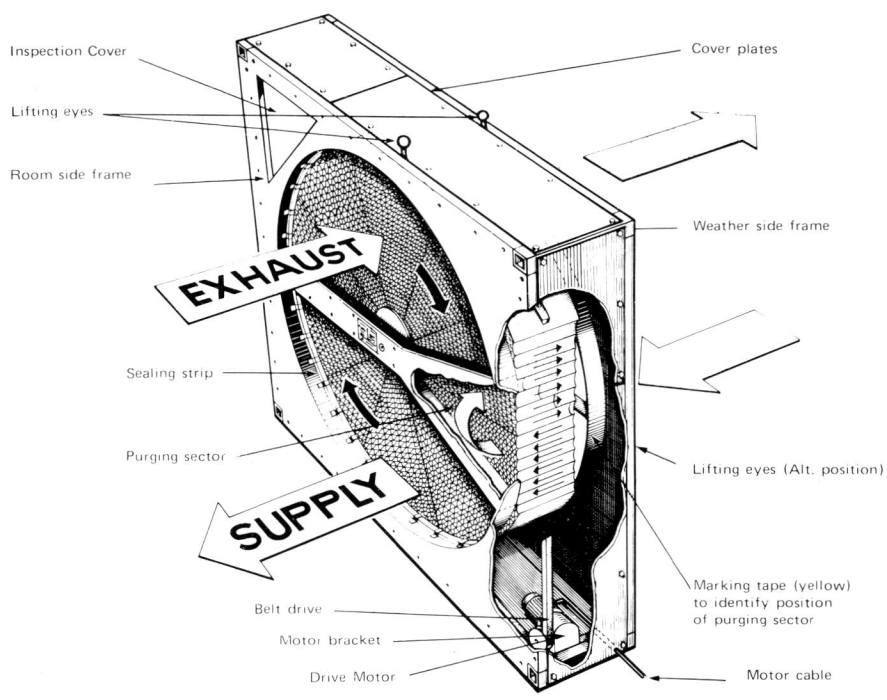

Inspection Cover

Lifting eyes

Room side frame

EXHAUST

Sealing strip

Purging sector

SUPPLY

Belt drive

Motor bracket

Drive Motor

Cover plates

Weather side frame

Lifting eyes (Alt. position)

Marking tape (yellow)
to identify position
of purging sector

Motor cable

*8.1 (above) The Econovent heat
wheel*

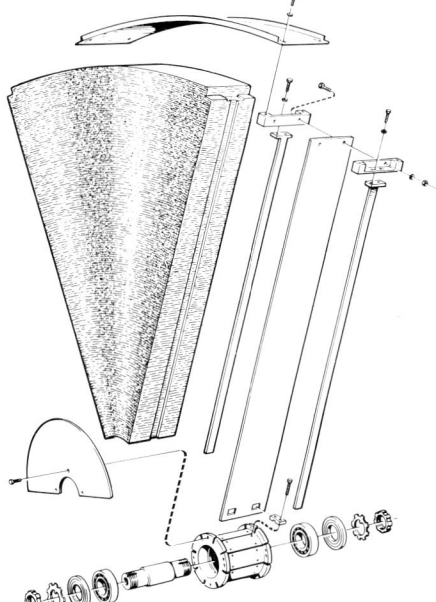

*8.2 (left) Design details of
Econovent heat wheel*

Carl Munter, also of Sweden, had the idea around 1959 to use Ljung-strom's principle to reclaim heat from the exhaust air of buildings. For some years such regenerators were called Munter's wheels, although of course Carl Munter was not really the inventor of the principle. Today numerous firms market heat recovery wheels of all kinds.

Basic principles

The thermal wheel consists of a thick disc-shaped matrix, made from metal or other material, which permits gases to pass through its thick-ness. It revolves slowly around its axis, driven by a low-power electric motor. In the heat absorption part hot waste gases are pumped through the matrix of the wheel, heating up this section. As the wheel revolves, the heated sector passes through the purge section. In this, fresh air is blown through, to remove the last traces of exhaust air which may remain in the passages of the wheel matrix. The heated sector next pas-ses through the supply air section, where cold external air is blown through the hot wheel matrix. The heat stored in the matrix is given off to the incoming air. The transfer of heat from the hot but impure outlet gases to the cold but clean supply air is a continuous one. The principle of the heat wheel can be used for numerous different domestic, com-mercial and industrial purposes.

The various thermal wheels on today's market differ from each other in important design aspects. The following units have been studied by the author:

Econovent marketed by Acoustics & Environmentics Ltd. of Claygate, Surrey.

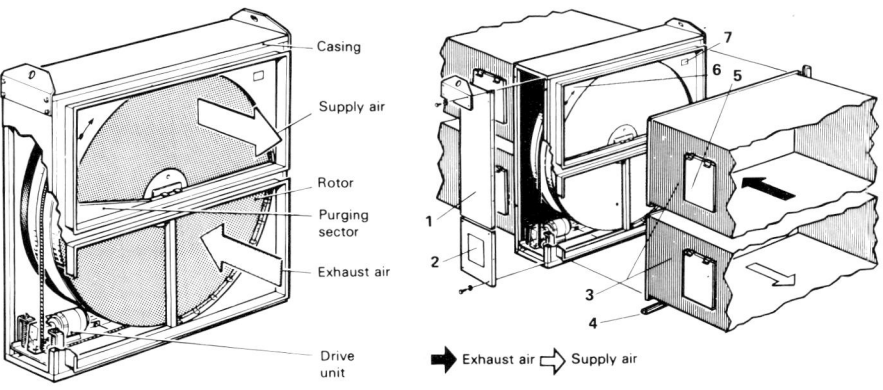

8.3 Flakt Regotherm rotary heat exchanger
  Exhaust air   Supply air
  1 Cover
  2 Inspection cover with plate, showing
    alternative methods of installation
3 Duct with PG joints
4 Guide rail
5 Inspection cover in duct
6 Duct connection instruction plate

The Bahco wheel marketed by Bahco AB of Sweden.
The Correx Wing marketed by James Heal and Co. of Halifax.
Regotherm marketed by Flakt AB of Sweden.
The Howden unit marketed by James Howden Ltd. of Glasgow.
The CN Heat regenerator made and marketed by Curwen and New-berry Ltd. of Westbury, Wilts.

The heat wheel units can rotate either on a horizontal axis, which appears to be the preferred solution, or on a vertical axis. Sizes vary between about 20 m^3 of air per hour for a small dwelling up to a maximum of around 100 000 m^3 of air per hour, which would correspond to a large factory. Small units have a wheel of only about 600 mm diameter, and perhaps only 50 mm thick. The biggest units at present on the market have wheel diameters of over 4 m with a wheel thickness of up to 250 mm.

Wheel construction

Different companies use differing methods of constructing the slowly rotating thermal wheel. The main criteria of construction must be:
Strength and durability
High thermal storage capacity
Ease of heat transfer with a minimum of pressure drop of exhaust gases and supply air
Correct thermal resistance design depending upon temperature of heat supply gas used
Corrosion resistance

The Econovent company employs fibrous inorganic asbestos paper in the corrugated form, which is rolled until a wheel of the desired dimensions is formed and there are a multitude of axial flutes or air passages, between 1 and 1.5 mm in diameter. After construction, the wheel is heated to drive the water of crystallization from the asbestos. It is then treated with lithium chloride solution to roughen the surface, thereby increasing the boundary layer heat transfer, as well as hardening the surface. The matrix wheel has a hard central core forming the axle which runs in ball bearing races at each end. A small electric motor keeps the wheel running at a speed not exceeding 10 rpm. The power consumption of the motors driving all thermal wheels is very modest. Even the largest thermal wheels ever built are rotated by electric motors with a consumption of well below 750 W. The smaller units with diameters of less then 2 m only need 250 W for the rotation of the matrix wheel.

The Econovent system is basically intended for heat recovery from exhaust air rather than for high temperature applications. There is an efficient purge section to prevent the carry-over from the exhaust flow into the supply flow. This carry-over has been found in all cases to amount to less than 0.04 per cent by volume. It makes the system an ideal one for hospitals, as there is little danger of bacteria being transferred from the exhaust air into the fresh air supply.

Bahco units are very similar in design, except that the wheels consist of corrugated aluminium in conjunction with corrugated asbestos, but the aperture diameter is the same: 1 mm to 1.5 mm.

Flow of gases through both Econovent and Bahco units is entirely streamline, and depends upon the roughness of the surface for efficient heat transfer. It is essential to use either of these systems under conditions where condensation does not occur, or is unlikely to do so. The systems are also totally unsuitable for use in environments where fluffy particles are suspended in the air, as for example in textile factories. There is then considerable danger that the narrow orifices may get clogged up, particularly in view of their rough surfaces.

The Wing Correx and Regotherm heat wheels employ a corrugated rotor made from aluminium sheeting. Such units can be used with exhaust gases having temperatures of up to 150°C. As the surface of the aluminium is quite smooth, and the internal width of the flutes is between 8 mm and 15 mm, the flutes are far less likely to get clogged up than than those in corrugated asbestos construction.

Regotherm use two different spacings, the wider one being employed for industrial uses in which there is some danger of blockage occurring. The Wing Correx is particularly suitable for operation with exhaust gases which contain suspended materials of a clogging nature, such as fibres.

The Howden units use wheels made from a series of corrugated metal sheets, the design of the corrugations varying according to the use for which the system is intended. The heat wheel is constructed from a number of such modular units. These are vitreous enamelled to afford good chemical resistance to flue gases, which are heavily laden with corrosives from the combustion of high sulphur fuels. The Howden regenerators are mainly intended for heat recovery from very heavily polluted industrial waste gases.

The heat exchange medium used with the CN heat regenerator consists of either aluminium or stainless steel knitted wire mesh, built up in sectors. Up to 24 sectors are assembled into each of the rotor wheels, which are made from aluminium, mild steel or stainless steel sheeting, depending upon the intended application. Each sector is effectively gastight in the plane of rotation.

Normally, where the design incorporates corrugated ducts it is essential to use heat wheels under conditions where no condensation can take place within the wheel. This does not matter with the CN design. One can operate the CN unit at dewpoint because condensate can drain away, helped by the centrifugal forces acting upon materials entrained in the wheel. In fact such condensate is fairly useful too, because it acts as a cleansing agent for the interstices between the knitted mesh. Heat

8.4 Curwen and Newbery heat wheel

transfer is actually aided by condensation occurring in this way. Stainless steel units are used for conditions where the exhaust gases are corrosive, while aluminium mesh is employed for more general purposes.

The degree of cross-contamination with CN units is, on the other hand, a good deal higher than with thermal wheels which use corrugated sheet materials in their construction. While the amount of cross-contamination for the Econovent has been shown to be 0.04 per cent by volume or less, the CN company only claims a limit of cross-contamination of about 1 per cent for their units.

The CN heat regenerator can be used to reduce the water vapour content of the exhaust by operation at dewpoint. Alternatively it can be fitted with a special purge unit to reduce dust and particulate discharge, as a side benefit to heat recovery. In some cases it is even possible to use these recovered materials for commercial purposes.

The following maximum temperatures are specified for CN units:

Aluminium media and aluminium rotor units 204°C
Stainless steel media and mild steel rotor units 427°C

THE USE OF FLUE GAS AS AN ENERGY SOURCE

When a hydrocarbon burns one can express this combustion chemically as:

$$C_n H_m + (n + M/4) O_2 \rightarrow n CO_2 + M/2 H_2O$$

For example, let us assume that we are burning fuel oil with a stoichiometric formula of $C_{20} H_{42}$.
One kg mole of this fuel oil weighs
$((20 \times 12) + 42) = 282$ kg
and would produce $20 \times 44 = 880$ kg of carbon dioxide and

$$\frac{42 \times 18}{2} = 378 \text{ kg of water vapour.}$$

Each kg of oxygen is associated with 3.262 kg of nitrogen, and in normal furnace operation one uses an excess of about 25 per cent combustion air.

This means that the amount of free oxygen in the flue gas would be equal to:

$$0.25 \left(20 + \left(\frac{42}{4}\right)\right) \times 32 = 244 \text{ kg of oxygen and}$$

$$1.25 \times 3.262 \left(20 + \left(\frac{42}{4}\right)\right) \times 28 = 3482 \text{ kg of nitrogen}$$

To sum up, we can now calculate the total mass of flue gases evolved:

Carbon dioxide	880 kg
Nitrogen	3482 kg
Water vapour	378 kg
Oxygen	244 kg
Total:	4984 kg

Lime kiln

8.5 Heat recovery from a lime kiln. Courtesy N. B. Orr, James Howden Ltd

It can therefore be seen that the combustion of 282 kg of fuel oil produces 4984 kg of flue gas. 17.67 kg of flue gas are thus liberated per kg of fuel oil burned. Obviously, if different fuels are used, the flue gas quantities are different. Equally, they differ when a lesser or higher excess of combustion air is employed.

The specific heat of flue gas varies according to its temperature. The specific heats of the constituents of flue gas, at 20°C and at constant pressure, are as follows:

Oxygen   0.917 kJ/kg K
Nitrogen   1.01   kJ/kg K
Carbon dioxide   0.838 kJ/kg K
Water vapour   4.181 kJ/kg K

The approximate specific heat at constant pressure of the flue gas mix is thus 1.2125 kJ/kg K at 20°C. At 100°C the specific heat equals 1.2247 kJ/kg K, while at 200°C it is equal to 1.249 kJ/kg K rising to 1.26 kJ/kg K at 500°C.

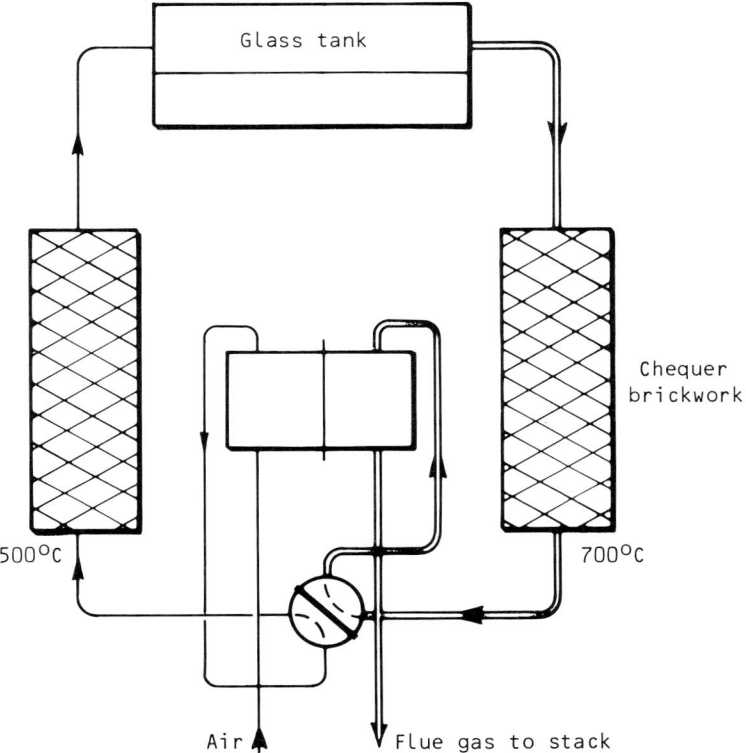

8.6 Heat recovery from a flat glass tank. Courtesy N. B. Orr, James Howden Ltd

The actual heat which can be extracted from flue gas is equal to:

$$17.67 \, W \times C_{ave} \times (t_f - t_e) \text{ kJ/day} \qquad (8.1)$$

where W is the mass of oil burned per 24 hour day in kg

C_{ave} is the average specific heat of flue gas between the temperatures at which heat is extracted in kJ/kg K

t_f is the temperature of the flue gas as it enters the thermal wheel or other heat exchange device in °C

t_e is the temperature at which the flue gas leaves the thermal wheel or other heat exchange device, again in °C.

For example, let us consider a plant where 1 tonne (1000 kg) of fuel oil is burned per day. The flue gases leave the waste heat boilers at a temperature of 200°C, to be cooled down within the thermal wheel to 60°C before being discharged to the atmosphere.

The average specific heat between 60°C and 200°C of the flue gas works out at 1.23 kJ/kg K.

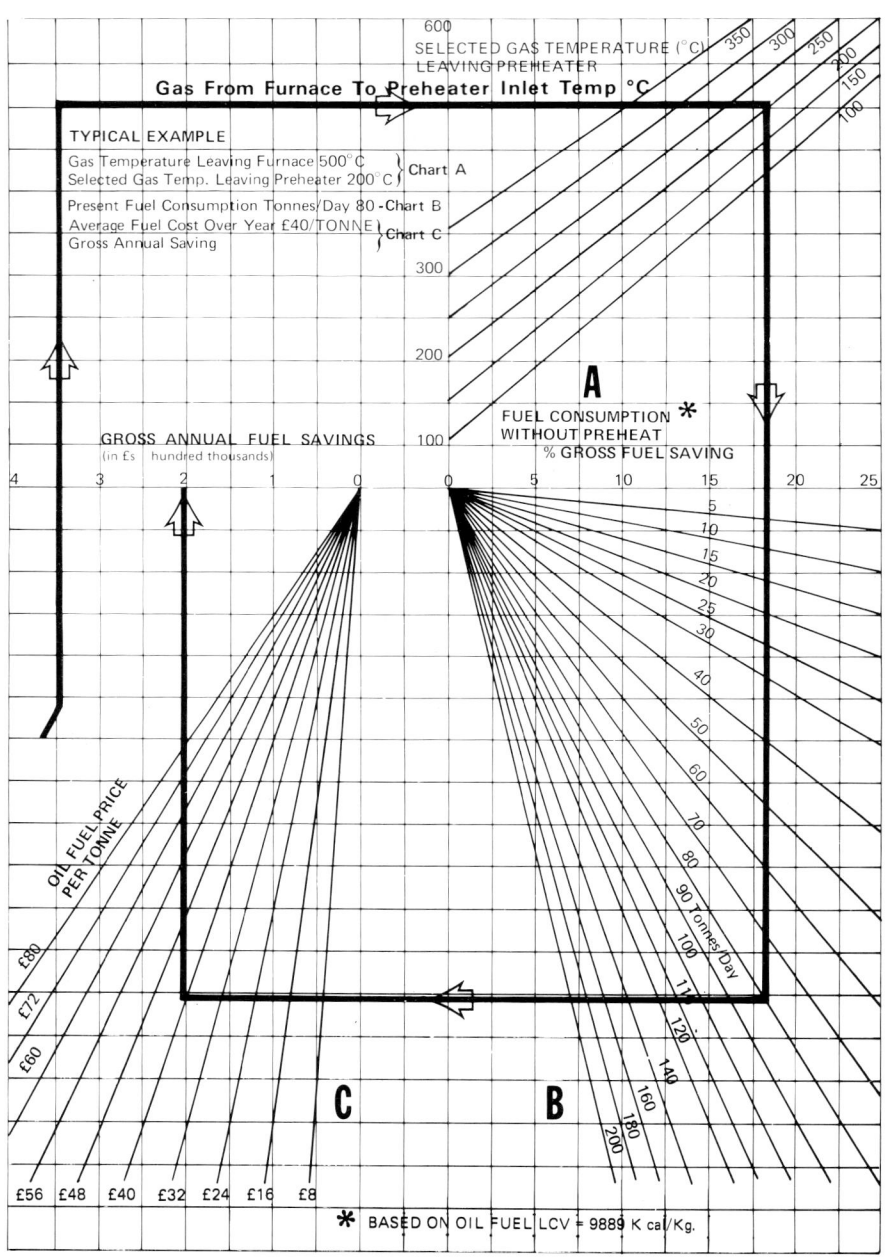

8.7 Nomogram determining savings with Howden heat wheel

The amount of heat recoverable is therefore:
$17.67 \times 1000 \times 1.23 \times (200 - 60) = 3\,042\,280$ kJ/day = 3042.28 MJ per day

Much of this heat can be used for preheating combustion air. The lowering of the flue gas temperature in a furnace reduces heat losses very drastically.

An experment carried out by James Howden Ltd. involved a moderately rated furnace which burned 90 t of heavy fuel oil per day, providing flue gas at 500°C. Lowering the flue gas temperature to 200°C produced fuel savings of around 18.2 per cent. As £84 per tonne was paid at the time for fuel oil, the annual fuel savings could be calculated at £420 000. The sensible heat obtained when abstracting the waste heat from flue gases can be used for heating up incoming fresh air.

Preheating combustion air
The sensible heat in the flue gases can be employed to preheat the air used for the combustion process and to enhance the efficiency of combustion of the fuel in this way.

This has the following positive effects:
a Higher flame temperatures ensure more complete combustion
b It is possible to operate with reduced levels of excess air
c Burner maintenance is reduced

To counteract the improved efficiency of combustion it becomes possible to cut back on fuel feed. When reducing the stack temperature from, say, 500°C to 180°C by the installation of a regenerative heat exchanger, less fuel and less excess air are used. The flue gas velocity is lowered drastically from its normal rate of about 7.5 m/s. It is usually necessary to introduce a forced draught flue gas system and to reduce the diameter of the chimney.

Most forced draught systems operate with a minimum efflux velocity of about 15 m/s. The use of a regenerative heat extractor together with combustion air preheating reduces the flue gas efflux to about half. Replacement of a natural draught system with cold combustion air in which the flue gases travel at 7.5 m/s by a forced draught hot combustion air system in which the velocity of the flue gases is 15 m/s calls for a halving of the chimney diameter. The reason for this is that the velocity of throughput is effectively reduced to a quarter.

Economic evaluation of combustion air pre-heating
This study was carried out by Norman B. Orr of James Howden and Co Ltd of Glasgow. The costs are evaluated at 1980 price levels. The plant studied is a crude distillation unit in an oil refinery consuming 90 tonnes of heavy fuel oil per day. This has a calorific value of 41.4 MJ/kg. The furnace gases enter the pre-heater at 496°C, and the stack at 180°C thus pre-heating air at 10°C to a temperature of 343°C.

Fuel cost	£84 per tonne
Rate of fuel saving	20.1 per cent

Total annual fuel saving $\dfrac{£84 \times 0.201 \times 90 \times 8000}{24} = £506\,520$

(assuming 8000 h/y of operation)	
Increase in annual maintenance costs	£3 960
Increase in fan running costs	£5 729
Actual net savings of energy per annum:	£496 831

Capital costs:	
Cost of pre-heater, including erection and site positioning	£93 364
Cost of insulated ductwork plus erection	£77 722
Capital cost of fans plus positioning	£14 797
Capital cost of fan motors including erection	£6 492
Capital cost of new Y-jet burners	£28 116
Cost of furnace/burner modification	£9 412
Total cost of fitting annular cap to chimney	£1 882
Total capital cost	£231 785

Break-even period: $\dfrac{£231\,785}{£496\,831}$ years = approximately one half-year.

INSTALLATION OF ROTARY HEAT REGENERATORS

It is possible to position rotary heat regenerators either horizontally or vertically. It is necessary to instal fans so that:

a Supply air is pumped in after it has passed through the heat regenerator wheel, ie on the inside of the building.

b The exhaust air is pumped out after it has passed through the heat regenerator wheel, ie on the outside of the building.

There are two exceptions to this:

1 If it is essential to eliminate the risk of equipment room air entering the supply, a pumping system is installed on the outside of the building. Air is pumped in under considerable pressure, thereby over-pressurizing the supply.

2 Where long and pressure-demanding systems have to be supplied with air, pumps are used both outside and inside the feed system.

In order to obtain effective purging, the static pressure of the supply air before it reaches the heat regenerator wheel must always be at least 360 pascals higher than that of the exhaust air which has passed the wheel. An excessive pressure difference, however, causes too much purging. In general, the ratio

$\dfrac{P_3}{P_2}$ should be less than $\dfrac{2}{3}$

It is always necessary to achieve a reasonable balance of supply and exhaust air volumes. If these differ from each other by more than 20 per cent it may be necessary to instal a bypass duct.

Filters

Filters inevitably increase the pressure requirements of the fans. Normally inlet air enters through an intake louvre and passes through a 1 mm fine mesh sieve. In areas where air is reasonably clean no further filtration is necessary; in town centres, industrial areas and other locations with heavy air pollution it is advisable to include a filter in the duct which leads from the outside to the wheel. Filters on the inside of the supply air ducts vary according to the cleanliness requirements of the rooms to be serviced.

No filter is normally fixed to the exhaust air from hospitals, schools, hotels and the like. It is, however, necessary to protect the thermal wheel when exhausting industrial gases, particularly when large particles of greasy, sticky or baking particles and dust are involved.

The efficiency of the heat recovery wheel increases rapidly with an increase in the number of revolutions per minute: at 4 rpm the efficiency of the wheel in removal and re-supply of sensible heat has reached about 78 per cent. This goes up to a mere 80 per cent when revolving the wheel at 10 rpm. Hence, an increase of speed beyond 4 rpm only provides very marginal improvements in overall efficiency

Use of roof-top exhaust heat recovery systems for pre-heating fresh air

The Bahco AEL system is a composite unit with two rotary heat regenerators, pumps and filtration equipment for reclaiming heat from exhaust air. The modular packaged unit is approximately 10 m long, 4 m wide and 2.5 m high and has a light metal framework with weatherproof cladding.

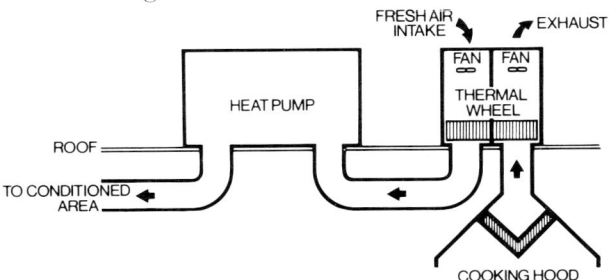

8.8 Combination of heat wheel and heat pump for energy conservation. Courtesy Electricity Council

The company makes six basic variants in design of the system. Exhaust air is sucked in from the building and gives off its sensible heat to twin rotary heat regenerators before being expelled to the outside. External air is heated by the waste heat abstracted from the exhaust gases. It then passes through a fairly elaborate filtration plant before being pumped into the building through the inlet duct. If necessary, additional heat is added to the inlet air by passing it through a standard run-around liquid/gas coiled heat exchanger. All control equipment that forms part of the factory-built package is within the unit. Every component access door is marked for easy identification, and there is enough space inside the unit to make it easy and comfortable to service.

The thermal wheels used with the AEL are made from corrugated aluminium and have a small central purging section. Before a section enters the inlet, some of the feed air is used to blow through the aluminium ducts for purging to the outside residual exhaust air in the channels of the wheel. The filter area is unusually large, and is designed for 9250 hours use; its filters system consists of standard cassettes measuring either 60 cm × 60 cm or 60 cm × 30 cm. The average pressure difference at the filter is about 100 pascals rising to a maximum of 250 pascals at 9 000 hours use. Capacities of the roof units vary between 0.5 m^3/s and 30 m^3/s.

Cycle 1: Warm exhaust air extracted from the building. The heat of the air is absorbed by stack A.

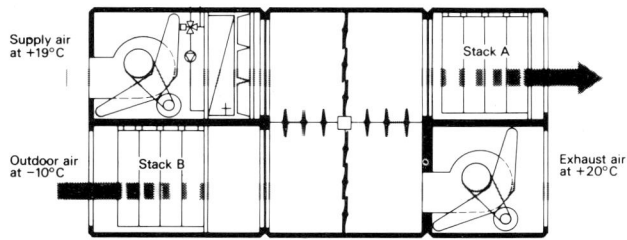

Patents subject to aluminium coils

Cycle 2: The damper has changed over. Cold air now flows through stack A and is heated.

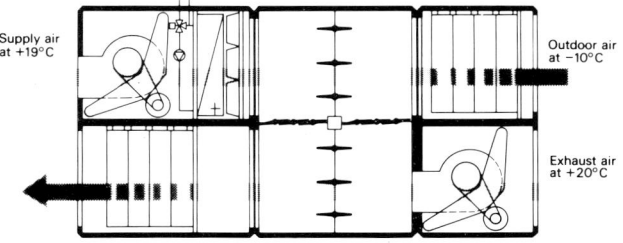

8.9 Operation of the Saphair heat regenerator system. Courtesy Saphair Ltd

The Saphair Resolair heat recovery unit
This heat regenerator is used to recover sensible heat from the exhaust air and transfer it to incoming fresh air. The system uses two accumulator stacks as heat absorbing units, made from corrugated aluminium alloy sheeting. The supply and exhaust air streams pass through these accumulator units alternately. In this way heat is alternately stored and given up. A damper unit is fitted between the accumulator stacks. The fans are connected either directly to the damper unit or across filters, heating coils, etc.

At any one time cold external air flows through one of the accumulator stacks, which has previously been pre-heated by exhaust air. Meanwhile, the warm exhaust air is allowed to heat up the other accumulator stack before being discharged to the outside. Each minute the air flows are reversed by the damper. The heat from the exhaust air is thus stored in one stack while the heat accumulated during the previous one-minute cycle is given off to the incoming supply air. The dampers which reverse the direction of air flow are reset by a motor which is operated electronically.

The temperature efficiency of the unit varies between 85 and 95 per cent at outdoor temperatures between $+10°C$ and $-30°C$. The enthalpy efficiency lies between 76 and 84 per cent.

Applications
Two versions of a similar design, known as the Kantherm, have been made, one for large installations with a maximum ventilation air flow of $40\,000\,m^3/h$, and one for small houses. The domestic units can be made for an air flow as low as 14 l/s using accumulators units 1 m long by 100 mm wide and 100 mm high.

It is calculated that if condensation is to be avoided, the air flow through every section of a dwelling should be at least 0.35 l/s and m^2 of floor area. Let us assume that a dwelling with a floor area of $185\,m^2$ is being considered. The ventilation rate needed is equal to:
$185 \times 0.35 = 64.75$ l/s or $225\,m^3/h$ of air.
Assuming an indoor temperature of $22°C$ and a mean external air temperature of $9°C$ over the year, the ventilation heat losses amount to 19.8 GJ of heat per annum.

Let us now assume the heat is being produced at a cost of £5.56 per GJ: the actual cost of heat lost per annum due to ventilation is then equal to £110. A suitable heat regenerator system can save some 90 per cent of these heat losses, thereby providing fuel savings of £100 per annum.

Naturally, before one can consider the installation of such a system, it is necessary to have a house built to Continental European rather than British or even American quality standards: doors and windows must be well-fitted and airtight.

SELECTION CHART: HEAT REGENERATORS

Type of equipment	Advantages	Disadvantages	Appropriate applications
Corrugated ceramic heat wheels	High temperature resistance Excellent heat storage capacity per unit mass	Easily clogged Limited life Low thermal conductance	Abstraction of heat from very hot gases
Corrugated fluted metal heat wheels	Little likelihood of getting blocked up Little carry-over between gas streams (under 0.04%)	Cannot be used under dewpoint conditions	Used where carry-over between streams of gases is undesirable, because the hot gas is polluted
Knitted wire heat wheels	Can be used under condensation conditions Reasonably cheap	High carry-over between primary and secondary gas flows	For general domestic, residential, office accommodation as well as sports halls, indoor swimming pools
Accumulator stack heat regenerators	Excellent heat recovery Can be positioned unobtrusively	High cost Liability to attack from corrosive gases	For private homes and blocks of flats, offices, schools, hotels, where air flow is less than 10 m³/s.

References

This chapter was written entirely with the help of the following industrial organisations:

Bahco Ventilation Ltd, Beaumont Road, Banbury, Oxon OX16 7TB, UK.
Kantherm AB, S-730 60 Ramnas, Sweden.
Econovent: Acoustics and Environmentics Ltd, Ruxley Towers, Claygate, Esher, Surrey KT10 0UF, UK.
Curwen and Newberry Ltd, Westcroft Works, Alfred Street, Westbury, Wiltshire BA13 3DZ, UK.
AB Flakt, Fack S-104 60, Stockholm, Sweden.
James Howden and Company Ltd, 195 Scotland Street, Glasgow G5 8PJ, UK.
James H. Heal & Co. Ltd, Richmond Works, Halifax, West Yorkshire, UK.

SELECTED COMPANIES INVOLVED IN MANUFACTURE OF PRODUCTS

UK and Europe

Acoustics and Environmentics Ltd, Ruxley Towers, Claygate, Surrey KT10 0UF.

Bahco Ventilation Ltd, Beaumont Road, Banbury OX16 7TB.

Beltran Ltd, Sunderland Street, Macclesfield, Ches SK11 6JF.

F. H. Biddle Ltd, Newtown Road, Nuneaton, Warks CV11 4HP.

Curwen and Newbery Ltd, Westcroft House, Alfred Street, Westbury, Wilts BA13 3DZ.

Howden Group Ltd, 195 Scotland Street, Glasgow G5 8PJ.

Redwood Industrial Products Ltd, Unit 14, Hither Green Industrial Estate, Clevedon, Avon BS21 6XU.

Rossfor Associates Ltd, 37 Lowlands Road, Harrow, Middx HA1 3AW.

Saphair Ltd, 12 Market Hill, Southam, Warwickshire CV33 0HF.

S&P Coil Products Ltd, Evington Valley, Road, Leicester LE5 5LU.

Stone Platt Crawley Ltd, Gatwick Road, PO Box 5, Crawley, West Sussex RH10 2RN.

Strax Distribution Ltd, 41B Brecknock Road, London N7 0BT.

United States

Banson Co. PO Box 10458, Winston-Salem, NC 27108.

Basic Environmental Engineering Inc., 21 W. 161 Hill Avenue, Glen Ellyn, IL 60137.

Beltran Associates Inc., 200 Oak Drive, Syosset, NY 11791.

Energy Recovery Co. Div. of Wehr Corporation, 448 S. Main Street, Verona, WI 53593.

Cargocaire Engineering Corp., 79 Monroe Street, Amesbury, MA 01913.

Moli-Tron Co. Div. of Molitor Inc., 2829 S. Santa Fe Drive, Englewood, CO 80110.

Peabody Gordon Piatt Inc., PO Box 650, Winfield, KS 67156.

Pyradyne Inc., 300 Nichols, PO Box 190, Hutchins, TX 75141.

Thermodyne Inc., 506 Jasper Street, PO Box 1007, West Columbia, SC 29171.

J. L. Underwood Co., 4450 Commerce Drive, S.W., PO Box 43048, Atlanta, GA 30378.

United Air Specialists Inc., 4440 Creek Road, Cincinnati, OH 45242.

WeatherRite Division Acrometal Products, 2240 Terminal Road, Roseville, MN 55113.

Wing Co. Div. of Wing Industries Inc., 125 Moen Avenue, Cranford, NJ 07016.

9   Heat pipes

In normal heat exchangers heat travels from one side of a thermal conductor to the other by simple conduction. The rate at which heat can be transferred is governed by the thermal conductivity of the interface material, and by the thermal resistances of the air/metal or fluid/metal interfaces. Thermal conductivities of metals are much higher than those of non-metals, amounting to 80.3 W/m K for steel, 398 for copper and 237 for aluminium.

In the heat pipe heat is transferred by convection and change of state. A liquid is evaporated at the absorber or evaporator end, the heat being abstracted in the form of latent heat. This is enormously larger in amount than sensible heat. For example, one requires about 420 kJ of energy in order to heat up 1 kg of water from its freezing point (0°C) to its boiling point (100°C). In order to convert 1 kg of water at its boiling point to steam at the same temperature, one needs a further 2258 kJ. Steam, being a gas, readily flows from one position to another, needing a negligible pressure head to do so. Once it comes to the other end of the heat pipe, all the latent heat is reclaimed as the steam condenses once more to water at the same temperature. If one then transports the condensed water back in some way to the evaporator section, one has a thermal siphon.

Thermal siphons have been known for a long time. In them heat which is taken in by a liquid at the lowest portion of a cylinder sealed at both ends, evaporates a liquid. The liquid vapourizes and its vapour travels to the upper end of the system. There it once more condenses to form a liquid. Its latent heat of liquefaction is given off and the fluid returns to its original position at the lowest level of the system by gravity, to start the cycle afresh. The limitation of simple devices of this type has always been that the absorbing end of such a thermal siphon had to be well below the condensing section.

In 1944 R. S. Gaugler took out a US patent on a thermal siphon system in which this limitation did not apply. It was not until about 1968—1970 that the term heat pipe was introduced, and that practical use was made of such devices.

As in the thermosiphon, the basis of the heat pipe is the evaporation of fluid at one end of the unit and the flow of the fluid vapour to the other, where the heat is regained by condensation of the fluid. The important task which has to be performed by a viable heat pipe is to

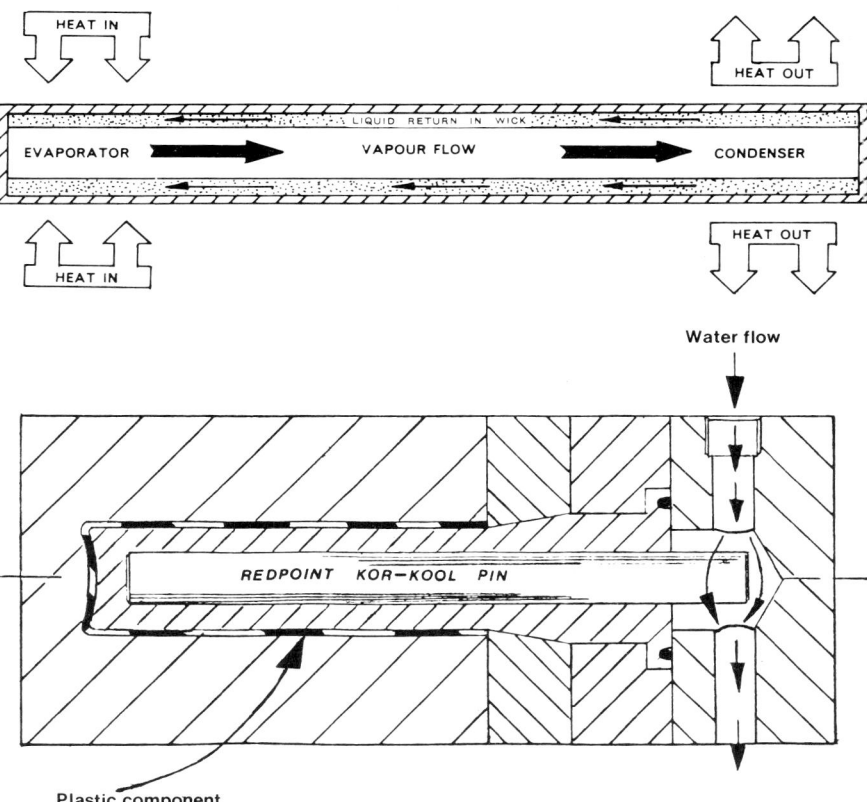

9.1 Schematic diagram and installed example of Redpoint Kor-kool heatpipe pin

transport the condensed liquid back to its starting position.

Although it is possible to use gravity, as with the traditional thermo-siphon units, modern heat pipes usually rely mainly upon capillarity, employing a so-called wick.

In practical designs the sealed metal tube is lined along its internal perimeter with a porous material which is easily wetted by the liquid employed. At the heating end the liquid in the wick evaporates, travelling along the hollow central portion of the heat pipe to the other end, where it gives off its heat as latent energy. The liquid condensate impregnates the porous material and passes back to the heat absorbing end of the heat pipe by capillarity. Although they perform better when the fluid flow is gravity-assisted, wicked heat pipes operate completely satisfactorily even when the heat absorbing portion lies slightly above the heat donating section.

One should, however, if at all possible, have the heat-absorbing eva-porator section below the heat donating condensing section. It has been found that, everything else being equal, such devices operate about 3–5 times as efficiently as other devices where the liquid flow is not gravity assisted. Properly designed heat pipes of this type are extremely efficient transporters of thermal energy. For example, a copper heat pipe tube with 9.5 mm external diameter and 300 mm in length, can transfer heat at the rate of 165 W if filled with water as the heat pipe medium. The only temperature drop between the two ends in approximately 2°C because of the need to maintain slight superheat to enable water to boil. Copper has one of the highest conductivities of any metal. Yet if one wished to maintain a heat transfer of 165 W with a copper rod of the same dimensions, one would have to produce a thermal difference between the two ends of the rod of over 1800°C. That, of course, would be totally impractical.

Sweat glands
As with many of man's clever devices, heat pipes are also widely found in nature. In man and other mammals, excess body heat is removed by sweat glands. Water is vapourized inside the body and heat is removed in the form of latent heat. The water vapour then condenses at the surface of the skin, being once more converted into water, thereby giving off the heat to the atmosphere. Some sweating causes the loss of body fluids to the atmosphere. It has now been found that most body cooling involves so-called eccrine glands in which the sweat is not lost but is conducted back towards the centre of the body by capillarity. In other words animal and human bodies use heat pipes to keep their temperature steady.

THE DESIGN OF HEAT PIPES

Heat transfer rate with heat pipe
The average heat pipe of 25 mm diameter can transfer about 2 kW of thermal energy from the heat input section to the heat output section. It has to be fitted on both input and output sides with heat exchangers which can cope with distances of 6 metres or more between these sections, provided the condenser is either at the same level as or above the evaporator. In some cases heat transfer between the evaporator and the condenser is controlled by varying the inclination of the heat pipes. If the condenser is below the evaporator, the flow of fluid is stopped and no heat is transferred. As the level of the condenser is raised relative to the evaporator till eventually the evaporator is below the condenser, heat transfer is improved because the rate of flow of fluid from condenser to evaporator increases. There are some designs of heat pipes with which it is possible to transfer heat from an upper level to a lower level, such as the osmotic heat pipe, the rotating heat pipe and some in which normal capillarity is augmented by some sort of external pumping

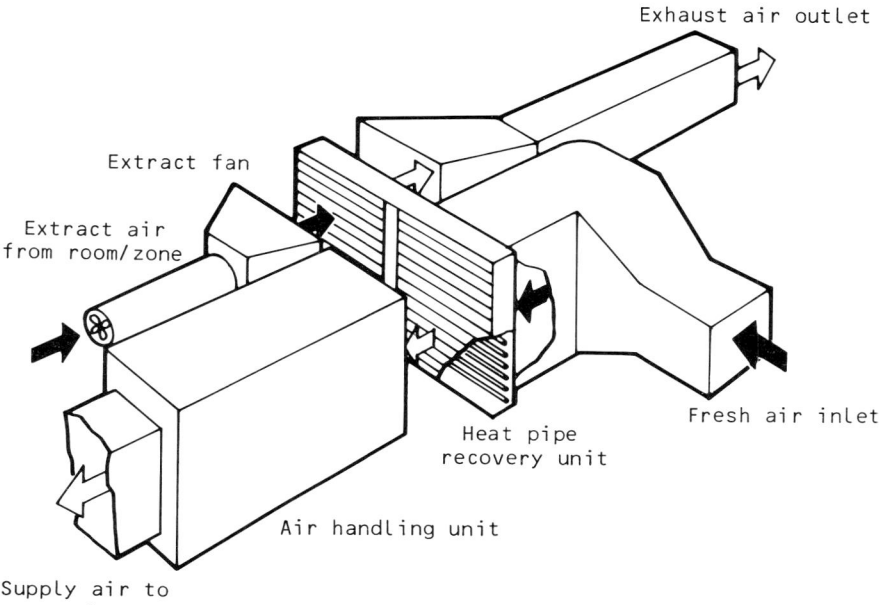

Exhaust air outlet

Extract fan

Extract air
from room/zone

Fresh air inlet

Heat pipe
recovery unit

Air handling unit

Supply air to
room/zone

9.2 Operation of heat pipe exchanger system. Courtesy Redpoint Ltd

arrangement such as electro-osmosis. These are not widely used, being restricted to certain limited and highly specialized applications.

For normal operation the **heat receiver should be either at the same level as or above the heat donator.**

Calculation of thermodynamic efficiency of energy transfer

The reason why a heat pipe is so very effective in transferring energy is that conduction and convection heat losses from the fluid in the pipe are minute in comparison to the thermal value of the heat transferred. The intrinsic thermodynamic efficiency of the process is extremely high. If we compare heat losses from the hot fluid to the outside, these would be of the same order of magnitude as would take place when heat is being transferred simply by a flowing hot liquid. Yet while 1 kg of hot water at 100°C represents a heat quantity of 40 × 4.186 = 167.44 kJ, assuming that during the take-off of heat it is cooled down to 60°C, the same quantity of water in terms of its latent heat represents 2258 kilojoules: ie 13.5 times as much thermal energy. In addition, the cooling down of water from 100°C to 60°C represents a considerable increase in entropy, which is a thermodynamic way of saying that one has lost a lot of high-grade heat.

The increase in entropy when water is cooled down from 100°C to 60°C equals 1310 − 831 = 479 J/kg K which gives a thermodynamic efficiency of heat transfer of

$$\frac{(1310 - 479) \times 100}{1310} = 63.44 \text{ per cent}$$

Heat transfer using flowing liquids is therefore accompanied by a considerable loss in energy grading: the energy received is less valuable from a thermodynamic point of view than the energy sent out.

In heat pipes, the difference between the grading of heat sent out and the heat received is far less. While heat is given out at the boiling point of the fluid, the temperature at which the fluid is made to boil is the boiling point of the fluid plus a slight elevation necessary to overcome vapour bubble effects. This is called the superheat of boiling and amounts to the following values for normal metal surfaces in contact with fluids used in heat pipes.

TABLE 9.1 BOILING POINT AND SUPERHEAT OF TYPICAL HEAT PIPE FLUIDS

Fluid	Boiling point (°C)	Average superheat (°C) needed for boiling
Nitrogen	− 196	0.3
Ammonia	− 33.5	2.0
Freon 11	24	1.8
Methanol	64.0	0.5
Ethanol	64.5	0.5
Water	100.0	2.0
Mercury	361	5.5
Potassium	773.9	9.0
Sodium	892.0	26.5
Lithium	1340	45.0

The change in entropy involved equals L/T_b when a fluid boils and $-L/T_c$ when the vapour is re-condensed

where L   is the molar latent heat in J-mole K

T_b is the effective boiling point of the liquid (including superheat) in kelvins

while T_c is the condensation point of the vapour which equals the theoretical boiling point (without superheat), again in kelvins.

When water is used as the heat pipe medium, one can write that the molar latent heat equals 40 644 J/mole while superheat amounts to 2.0°C.

The thermodynamic irreversibility can therefore be expressed as:

$$\frac{40\,644}{373.15} - \frac{40\,644}{375.15} = 0.5807 \text{ J/mole K}$$

If we compare the overall useful entropy change as being

$$\frac{40\,644}{373.15} = 108.92 \text{ J/mole K}$$

it becomes clear that the irreversibility of heat pipe heat transfer, using water as a medium, is only 0.53 per cent. Conversely, the thermodynamic efficiency is of the order of 99.5 per cent (excluding conduction and convection heat losses).

Thermodynamic efficiencies when using either methanol or ethanol are even better than this. Even when molten sodium is employed, which gives the poorest characteristics in this respect, a thermodynamic heat efficiency of around 97 per cent can be obtained. No other systems of heat transfer can even approach such a figure.

Factors affecting the rate of heat transport with a heat pipe
In any physical heat pipe an equilibrium has to be set up so that the rate of evaporation and condensation equals the rate of travel of the condensed fluid back to the evaporator section. It is this rate of fluid travel which governs the whole efficiency of the heat pipe. One can express the heat transfer rate as being equal to:

$L \times G \qquad$ J/s

where L is the latent heat of the fluid in J/kg and
$\qquad$ G is the mass flow rate of the fluid from the condenser section to the evaporator section in kg/s

The rate of fluid flow is determined by the following factors:

a The pressure drop caused by differences in pressure between the evaporator section and the condensing section. As the pressure in the condensing section is lower than that of the evaporator section this causes the vapour to flow along the pipe. At the same time this pressure difference also impedes the liquid flow, although the effect is not very large.

b Capillarity is the main effect causing the fluid to flow from the condensing end to the evaporator end of the heat pipe.

c The gravitational head depends upon whether the evaporator is at the same level as the condenser, in which case this effect is zero. If it is above the level of the condenser, the flow rate of fluid is impeded, and therefore the capacity of the heat pipe is lowered. Equally, if the evaporator section is below the level of the condensing section, gravity helps the fluid flow and therefore increases the heat transfer capacity of the heat pipe.

There are certain other methods of increasing the fluid flow rate and therefore the heat transfer capability of the heat pipe, as for example, the use of centrifugal force in a rotating heat pipe system, or the incorporation of a pumping device.

For most simple heat pipe systems a wick is all that is provided to enable the fluid to return from the condenser to the evaporator. A wick is basically a porous layer which can be considered to constitute a large number of tiny capillary tubes, side by side.

Elementary physics explains that a liquid rises in a capillary due to the wetting effect of the fluid upon the solid material: the following equation gives capillary rise in such tubes:

$$h = \frac{2\ \sigma\ \cos\ \theta}{D\ g\ r}$$

where h is the height of capillary rise in metres.
   σ is the surface tension of the liquid in N/m.
   θ is the angle of contact made by the fluid with the surface of
      the solid material, in degrees.
   D is the density of fluid in kg/m^3.
   r is the radius of the capillary in m.
   g is acceleration due to gravity in m/s^2.

When a liquid wets a surface completely, as is the case when one uses hydrophilic fluids such as water, alcohol or ether against ceramic or similar surfaces, θ becomes 180°C so that cos θ approaches 1.

In a wick the capillaries are usually considered to be arranged horizontally, so that gravitational effects do not come into play, except in the circumstances already mentioned. The fluids flows through these capillaries in strictly streamline motion because Reynolds' number for the flow, expressed as:

$$\frac{v\ d\ D}{\mu}$$

where v is the velocity of flow in m/s
   d is the effective diameter of each capillary in m
   D is the density of the fluid in kg/m^3
   μ is the viscosity of the fluid in kg/m s

is always far below 2000, which is the transition point between streamline and turbulent motion. The driving force for the flow of the fluid in the wick is surface tension.

It is almost impossible to work out the action of a specific wick because of the way the channels are twisted and the uncertainties of measuring capillary pores. A useful practical equation to evaluate the travel of different fluids in wicks of different dimensions is Darcy's equation, which applies to horizontal systems.

This can be simply stated as:

$$G = \frac{K\ \sigma\ \mu\ A}{D\ l}$$

where G is the mass flow rate of the fluid along the wick in kg/s.
   σ is the surface tension of the fluid against the wick material in N/m.
   μ is the viscosity of the fluid at the temperature in question in kg/m s.
   A is the cross-sectional area of the wick in m².
   D is the density of the fluid in kg/m³ at the temperature concerned.
   l is the length of the wick in m.
   K is the constant applying to the wick material (dimensionless).

It can be seen that the rate of flow of fluid along a heat pipe is directly proportional to the cross-sectional area of the wick and inversely proportional to its length.

As the rate of heat transfer in a given heat pipe is limited by G × L where L is the latent heat of evaporation of the fluid in terms of J/kg, it can be seen that the maximum capacity of any given horizontal heat pipe is expressed as:

$$\text{Maximum heat transfer}: = \frac{K \, \sigma \, \mu \, A \, L}{D \, l} \quad \text{J/s}$$

Table 9.2 gives the values for σ, μ and D for a number of fluids at their boiling points.

TABLE 9.2 VALUES FOR SURFACE TENSION σ, VISCOSITY μ AND DENSITY D OF FLUIDS USED COMMONLY FOR HEAT PIPE PURPOSES AT THEIR BOILING POINTS

Fluid	Surface tension σ N/m	Viscosity μ kg/m s	Density kg/m³
Ammonia	2.8×10^{-4}	4.75×10^{-4}	771.0
Methanol	2.014×10^{-4}	4.03×10^{-4}	763.5
Ethanol	2.004×10^{-4}	5.04×10^{-4}	742.8
Diethyl ether	1.347×10^{-4}	2.33×10^{-4}	736.0
Water	5.89×10^{-4}	2.818×10^{-4}	958.4
Sodium	1.91×10^{-3}	6.9×10^{-4}	1079.0
Potassium	1.01×10^{-3}	5.6×10^{-4}	1206.2

Numerous other fluids such as the entire range of fluoro-hydrocarbons, toluene, and diphenyl are also commonly used in heat pipes, as well as helium, molten lithium and even molten silver for extreme use purposes. Materials of construction depend upon the fluids used, but normally include aluminium, copper and stainless steel.

Wick design

The wick for the return flow of the fluid from the condenser end to the evaporator end of each heat pipe is chosen primarily for adequate wettability and compatability with the fluid to be transported. The smaller the pore size, the better the capillary head exerted, but the lower the rate of fluid flow will be. For this reason low performance wicks in horizontal and gravity-assisted heat pipes should have fairly large pores, of the order of 50–100 μ m in radius.

When the fluid has to be pumped against gravity, one needs very small pores to obtain very good capillary pressure, but the flow of fluid through such a wick is very slow and the capacity of these heat pipes is small.

Thick wicks can transport more fluid and therefore increase the capacity of a given heat pipe unless the thickness of the wick militates against the easy evaporation of the fluid at the evaporator end.

Types of wicks used

Nickel is used in the form of powder, felt and foam. In the form of powder the capillary height capable of being overcome may exceed 400 mm, but for the other materials capillary pressures are low. Nickel felts are used mainly with heat pipes filled with water, methanol and other organic fluids.

Copper is used as a powder, where pore sizes can be as small as 10 μm in radius, with a resulting capillary height exceeding 1.5 m. The thermal capacity is poor because of slow fluid flow. Copper can also be used as a foam for short, high capacity heat pipes. Like nickel, it is used for water and organic fluids.

Stainless steel in the form of mesh or twill can be made in pore sizes between 30 and 130 μm. It is used in contact with gases such as helium, nitrogen and ammonia for cryogenic heat pipes, as well as liquid metals such as mercury, sodium and potassium.

In many cases there is a fluid flow area between the wick and the tube wall, which can consist of grooves or ducts.

Nature of fluids to be used in a heat pipe

Fluids used in a heat pipe must:

a Be chemically stable over long periods of time.

b Be easy to purify and degassify.

c Be reasonably cheap.

d Not react with the materials of construction of the heat pipe wick.

e Boil at the approximate temperature of heat input and delivery without requiring the heat pipe to be pressurized excessively, as this would involve a considerable and disadvantageous increase in the thickness of tube walls.

For normal temperature ranges the following fluids are used:

TABLE 9.3 HEAT PIPE FLUIDS

Fluid	Boiling point at 1.013 bars pressure (°C)
Dimethyl ether	− 23
$CHCl_2F$ (Freon 21)	8.9
CCl_3F (Freon 11)	23.8
Diethyl ether	34.5
$C_2Cl_3F_3$ (Freon 113)	47.6
Acetone	56.2
Methanol	65.0
Ethanol	78.5
Benzene	80.1
Cyclohexane	80.7
$C_2Cl_4F_2$ (Freon 112)	92.8
Heptane	98.4
Water	100.0
Toluene	110.6
Flutec PP9 (proprietary fluid)	about 170
Thermex (proprietary fluid)	about 270

For cryogenic purposes one uses gases at or near their liquefaction (boiling) points.

TABLE 9.4 GASES USED FOR CRYOGENIC HEAT PIPES

Gas	Boiling point at 1.1013 bars
Helium	−268.6°C
Nitrogen	−195.8°C
Ammonia	− 33.35°C

For very high temperature purposes heat take-off, liquid metals are used:

TABLE 9.5 METALS USED AS FLUIDS IN HEAT PIPES

Metals	Boiling point at 1.013 bars
Mercury	356.6°C
Potassium	774°C
Sodium	892°C
Lithium	1317°C
Silver	2212°C

Heat recovery from exhaust air

An important feature relating the choice of a specific kind of heat pipe heat exchanger is its recovery factor: this is determined by the following equation, when air flows are equal:

$$R = \frac{T_{sl} - T_{se}}{T_{ee} - T_{se}}$$

where T_{sl} is the temperature at which the supply air leaves the heat exchanger in °C.

T_{se} is the temperature at which the supply air enters the heat exchanger in °C.

T_{ee} is the temperature at which the exhaust air enters the heat exchanger.

The specific heat of air at constant pressure is 1.01 kJ/kg K. Let us assume that a building of dimensions 50 m × 30 m × 20 m = 30 000 m³ is to be supplied with heated air at 20°C and that a ventilation rate of 1.5 air changes per hour is needed. This means that 45 000 m³ of air have to be supplied each hour, or 12.5 m³/s of air, weighing 15 kg at 20°C, have to be pumped in.

Let us now assume that the average temperature of incoming air is equal to 5°C over an operating period of 3000 hours per year. It is therefore necessary to supply a total of 15 × 15 × 1.01 kJ/s to the incoming air = 227.25 kJ/s (kW) or 227.25 × 3000 = 681 750 kWh over the entire annual operating period.

If £ C_{kWh} is the cost of a kilowatt hour of fuel, the annual saving by installing a heat pipe with a recovery factor of R equals:
681 750 × C_{kWh} × R = annual heat savings in £

To this one needs to add the savings in capital cost of being able to install a smaller heating plant or, in cases where air conditioning is used, a smaller cooling plant.

It is convenient to deduct these capital charges from those engendered by the installation of the heat recovery unit. The net capital cost of the heat recovery unit is thus its cost *minus* capital savings in heating or cooling plant.

Let the capital cost of the heat recovery unit be £C and let the saving in capital cost of installing a smaller heating (cooling) unit be £C'. The net cost of the heat recovery unit is therefore £(C − C').

If the sum of interest, depreciation and maintenance works out at p per cent of the capital investment, then the annual cost of the heat recovery unit equals:

$$£\frac{(C - C') \times p}{100}$$

Provided that this figure is less than the savings engendered, the installation of the heat recovery system is worth while.

Deciding on optimum size of recovery unit

The larger the heat recovery unit used, the lower the speed of fixed volumes of both exhaust gases and supply gases past the heat pipes. This improves the heat transfer via the pipes and therefore the recovery ratio R.

However, this increase in recovery ratio is by no means linear, and if the speed of gases past the heat pipe ends is reduced much below 2 m/s improvements become marginal. Capital costs, on the other hand, increase almost in a direct ratio with the size of the units. To choose the correct dimension of heat pipes it is necessary to obtain the manufacturer's figures for the recovery ratio guaranteed with varying air flows for the equipment on offer.

Typical variation of recovery ratio with flow rate

The heat transfer coefficient of the heat pipe itself is very large compared with the heat transfer from air to the heat pipe evaporator section or from the condenser section to the air. The duration of thermal contact of the air with these sections is therefore of great importance. However, at lower speeds the heat transfer coefficient between circulating air and surfaces falls. **Table 9.6** gives the relationship between the recovery ratio of a standard heat pipe recovery system in which air flow rate equals air return rate, at varying flow speeds, using finning at 60 per 100 mm.

TABLE 9.6 VARIATION OF RECOVERY RATIO WITH AIR FLOW RATE PAST HEAT PIPE UNIT

Flow speed past heat pipes m/s	Recovery ratio (dimensionless)
1.0	0.740
1.5	0.745
2.0	0.718
2.5	0.678
3.0	0.652
3.5	0.627
4.0	0.607
4.5	0.570
5.0	0.545

As can be seen, little improvement in recovery ratio is obtained at very low flow speeds because there is a tendency for the air flow to become streamline, with very much reduced heat transfer. Equally, at very high flow rates, the recovery ratio falls off very drastically. Naturally, with any specific kind of equipment, recovery ratio figures may differ from those given above.

Within the practical range of air flow velocities, when the volume passing remains constant, the recovery ratio varies with the area of the

9.3 (above) Packaged heat pipe unit. Note two heat pipe batteries in series; this simplifies cleaning via access doors in the ductwork. Courtesy Curwen and Newbery Ltd

9.4 (left) Flat plate fin heat pipe heat recovery unit. Courtesy Curwen and Newbery Ltd

heat exchanger system. The optimum air flow speed is chosen on the basis of the relative costs of heat and capital equipment. For example, during a period of high fuel cost and low interest charges, one would select a larger capacity (and more expensive) heat recovery unit than one would if conditions were reversed.

Effect of supply/exhaust flow ratios

Heat pipe recovery systems are much more effective when the exhaust flow is small compared with the supply flow. For this reason it is most advantageous to use systems in which one uses a comparatively small volume of exhaust gas (preferably at an elevated temperature), such as flue gas or ventilation air from laundries or industrial plant to heat up large quantities of incoming air.

Table 9.7 gives the heat recovery ratio when both flow and return gases are being pumped in at a speed of 2.5 m/s.

9.5 Heat pipe heat exchanger. Courtesy D. A. Reay, International Research and Development Ltd

TABLE 9.7 EFFECT OF MASS RATIO UPON RECOVERY RATIO

Mass ratio Supply air (dl) Exhaust gas	Heat recovery ratio (dimensionless)
1.0	0.678
1.2	0.740
1.4	0.778
1.6	0.815
1.8	0.834
2.0	0.865
2.5	0.888
3.0	0.910

When very high mass ratios are used, heat pipe heat recovery systems give excellent performance.

Frost control

The outflowing air is usually heavily laden with moisture. The incoming air, on the other hand, is dry. In winter, if the incoming air is below freezing point there is a danger that the exhaust air may be cooled to below 0°C, so that the water carried in the exhaust air freezes, and causes frost deposition in the exhaust duct. Under such circumstances it is necessary to make sure that the temperature of the exhaust air does not drop below the frost point. This is done by restricting the efficiency of the heat pipe exchanger unit. In practice this is achieved by tilting the system, so that gravity now works against the flow of condensed fluid from the condenser to the evaporator. The heat recovery ratio is dropped to:

$$R \times \left(\frac{T_{ee}}{T_{ee} - T_{ee}} \right)$$

where R   is the normal recovery ratio.

T_{ee} is the temperature at which the exhaust air enters the heat exchanger in °C.

T_{se} is the temperature at which the supply air enters the heat exchanger in °C.

For example, let us consider a system which works at a heat recovery ratio of 0.75 under normal conditions. We assume that the internal temperature, which equals the temperature of air entering the exhaust system = 20°C. If, for example, the temperature outside drops to −8°C then there is a danger of frost formation on the exhaust side if the system is being operated as before at maximum heat recovery ratio.

To eliminate this danger, it is now necessary to drop the heat recovery ratio from 0.75 to

$$0.75 \times \left(\frac{20}{20-(-8)} \right) = 0.536$$

By suitable tilt control the recovery ratio of the heat pipe system is reduced to about 0.53 so that there is now no possibility of the air in the exhaust system causing the exhaust duct to frost up. With many systems the tilt control is carried out completely automatically. As soon as a temperature sensor discovers that the temperature of the outside air has dropped below 0°C, a relay operates the tilt control to reduce the heat recovery ratio. This eliminates any risk of frost formation in the exhaust duct.

CLASSIFICATION OF HEAT PIPES

Heat pipes do not have to have the circular cross-section the name implies: various other shapes are also used. In addition, several methods are employed to speed the flow of the liquid from the condensing section back to the evaporator section. The speed of this flow is the capacity-determining factor for any given system. The basic types of heat pipe are the following:

a Multiple tube type capillary heat pipes

In these a number of tubes lead from the evaporator section to the condenser take-off section, capillarity being the only means of obtaining a return of the fluid. The heat transfer coefficient between an external forced air system and a fluid inside a heat exchanger is about 50 W/m^2 K, while internal heat exchange values of up to 10 000 W/m^2K are easily achieved with horizontal heat pipes. For this reason the extended surface area of the evaporator and condenser surfaces must be between 20 and 30 times the cross-sectional area of the heat pipes joining them. Circumferentially grooved wick types of heat pipe, using water as working fluid, are claimed to have heat transfer coefficients as high as 80 000 W/m^2 K.

b Gravity-induced fluid flow heat pipe

In these the wick is comparatively small, being mainly present to provide thermal insulation between liquid and vapour. The liquid flows from the condensing section to the evaporator section by gravity.

c Osmotic flow heat pipe

This type was developed by the US Energy Conversion Systems Inc, and works as follows: a solution of sugar in water is used as the working fluid. On the evaporator side heat acts upon this sugar solution, so that pure water vapour is driven off. This water vapour condenses on the condensing side, to give off its latent heat. The vapour section is separated from the liquid flow section by a semi-permeable membrane of cellulose. Osmotic pressure therefore develops between the sugar solution and the water produced during condensation. In consequence the water is fed into the sugar solution, being thus driven to the evaporator section where water is evaporated once more. The advantage of this system is that osmotic pressure is much greater than the simple capillary pressure which one relies on in normal heat pipes. With osmotic heat pipes it becomes feasible to place the condenser well below the evaporator, which is impossible with capillary heat pipes. On the other hand, flow rates of water through the membrane are low and therefore such heat pipes have a much poorer capacity.

d Electro-osmotic heat pipes

If one applies an electric field along the level of the heat pipe, the speed of the fluid along the wick is enhanced by the effect of electro-osmosis. It is necessary to use fluids with high dielectric properties. With some designs it has been found possible to pump fluids for distances of up to 500 mm against gravity, which is difficult to achieve with conventional heat pipes relying only on capillarity.

e Inverse thermosiphon

In these heat pipes the evaporator section is above the condensing section. There is also a tube which dips into the condensing sump into which a small auxiliary heater is fitted. As the vapour condenses in the lowest section, it enters the return tube. The auxiliary heater induces the formation of vapour bubbles in this liquid, so that the density of the liquid/vapour column is now less than the pure liquid which surrounds it. This difference in density helps return the liquid into the evaporator section.

f Heat plates

These are not heat pipes at all, but plates in which the meshed wicks alternate with vapour passages. As with normal heat *pipes* the evaporator section is on one side and the condenser section on the other. Flat heat plates are normally only used horizontally.

g Flexible heat pipes

In applications where vibration of either the evaporator section or the condenser section is encountered it may be difficult to fit a rigid heat pipe. RCA Corporation and Eastman Inc have developed heat pipes with flexible bellows to make it possible to transfer heat from one section to the other. The design of a wick able to remain operational under such circumstances presented difficulties, but the problem has now been solved.

h The rotating heat pipe

As has already been pointed out, the governing factor in the design of any heat pipe is the speed with which fluid passes from the condensing section to the evaporator section. The rotating heat pipe is an ingenious way of using centrifugal force to augment gravitational and capillary forces. The rotating heat pipe has the shape of a truncated cone, revolving on its axis, with the evaporator section being wider than the condensing section. As soon as the fluid condenses in the condensing section it becomes subject to centrifugal force and flows back to the evaporator section at a rate proportional to the angular momentum exerted.

It is possible to overcome gravity with such heat pipe designs, and the condensing section can be much lower than the evaporator section. Systems of this type are widely used for cooling rotors in electric motors and generators.

9.6 Flat heat pipe. Courtesy Noren Products

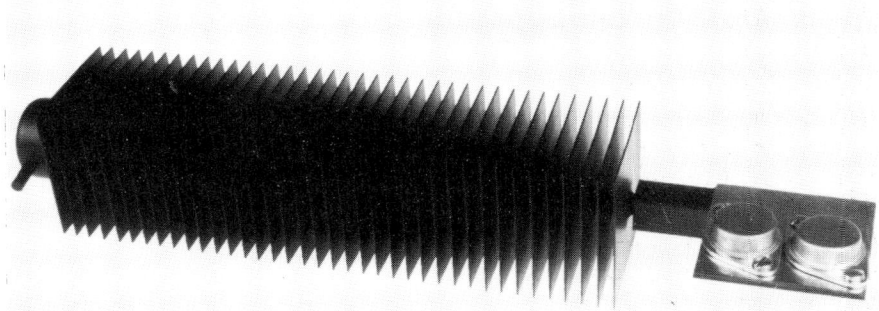

9.7 Variable conductance heat pipe. Courtesy Noren Products

Example of a proprietary heat pipe regenerator: the Q-pipe

Standard units consist of a number of heat pipes with nominal external diameter of 25.4 mm (1 in) usually made of aluminium alloy 3003-H14 or copper. Spacing of tubes is staggered, giving 57 mm c/c spacing transverse to flow direction, and 54.5 mm c/c spacing in the direction of flow. Tube walls are 1.245 mm thick. The tubes are finned at the ends with thin sheeting of the same metal as the tube itself, at a spacing of about 50 fins per 100 mm. A vertical partition between the two sections of the heat exchanger is provided, made from heavy gauge aluminium for alloy heat pipe systems and from stainless steel for the copper type. Casings are usually heavy gauge galvanized steel, although other materials can also be used in corrosive environments. The normal fluid used in Q-pipes is American National Standard Code R 12 at the rate of 290 g per m of heat pipe.

An important factor in the management of heat pipe units is tilt control. The system is designed so that units can be rotated in order to vary their inclination. If the evaporator or hot side is tilted so as to lie above the condenser or cold side the heat pipe capacity is modulated and reduced eventually to zero, thereby shutting down the heat exchanger. Tilt is used as an accurate control of the heat to be transferred.

In actual operation, the air flow rate recommended with Q-pipe units is between 1.78 and 3.5 m/s on both the supply and the return sides. If the two flows are within 20 per cent of each other, one should place the partition between the flows in the centre. Otherwise the partition should be fitted so that flow rates are roughly equal. Allowable limits of partition location are between 25 and 75 per cent of unit face lengths from one end.

PRACTICAL APPLICATIONS

Heat pipes are used as extremely effective heat exchangers for heat energy transfer at thermal levels which vary from the temperature of liquid helium, (boiling point 4.55 K) up to more than 2200°C, using liquid silver as a heat transfer medium. Because of the very remarkable efficiency of heat transfer from one end of the heat pipe to the other, and the fact that there is very little difference between the heat supply temperature and the heat take-off temperature, heat pipes are often employed for purposes where normal heat exchangers cannot be used.

Reclamation of flue gas heat

Very large quantities of sensible heat are normally lost to the environment in the form of exhaust gases from furnaces of all kinds. Waste heat boilers can be heated by gas/liquid heat exchangers in which circulating coils of water are passed through these exhaust gases. The problem with such heat exchangers is the very marked drop in temperature encountered. Because of the need to restrict the heat transfer area to a reasonable figure, it is necessary to circulate the water at a temperature much lower than that of the flue gases. Comparatively low pressure steam is therefore produced in the waste heat boilers. If one employs heat pipes, one can choose a fluid which boils at a temperature only a few degrees below that of the flue gases, and the waste heat can then be given off at almost the same temperature and be reused for such purposes as supply air pre-heating. As the flue gas has been cooled appreciably, it is now possible to introduce further heat pipe systems with fluids boiling at lower temperatures to recover lower grade sensible heat as well.

Heat pipe heat exchangers are used in many cases where heat has to be transferred between hot exit gases and cold inlet gases when there

is need to avoid mixing the gases, as may happen with rotary heat exchangers, even when purge conditions are good. As in normal heat
exchanger practice, the outflow and inlet gases are separated physically.

Heat pipes compete with heat recuperators in this field. They offer
the advantages of reduced flow resistance because heat absorber surfaces are far fewer, due to the much improved heat transfer characteristics of the heat pipe system. In general, one designs heat pipe heat
exchanger systems with an air flow of about 2–4 m/s past the neat
exchanger surfaces in order to keep pressure losses past tube bundles
fairly low.

Thermal recovery units in air conditioning systems

These are counterflow air-to-air heat exchanger units with the outside
appearance of a normal plate fin chilled water or steam coil. Such a
unit, however, differs from a normal heat exchanger in that each tube
which connects the two is a heat pipe.

Hot air is passed through one side of the exchanger and cold air
through the other in the opposite direction. A sealed partition separates
the two air flows to prevent any cross/contamination. It is possible to
recover between 60 and 80 per cent of the sensible heat difference
between the two airstreams. The same system can also be used to save
cooling costs in air conditioning systems. It is claimed that in such a
case, which gives almost the best return on expenditure, the cost of the
heat pipe heat exchanger system can be recovered by cost savings in
cooling equipment only. Units of this type are claimed to have the
advantages of being free from cross-contamination, more compact per
unit of performance, and needing negligible maintenance due to having
no moving parts. They can be applied for numerous purposes. The following are suggested:

a Use in industrial plant Wherever heated gases are exhausted to the
outside it is possible to use the sensible heat for pre-heating inlet air.
Due to the absolute separation of the two streams, incoming air is completely clean and pollution-free.

b Use in public buildings Heat recovery units are suggested for hospitals,
schools, factories, offices, hotels, department stores, auditoria and all
other kinds of similar buildings. The exhaust air, which may be laden
with cigarette smoke, food odours, germs and viruses is passed through
the evaporator section of the heat recovery unit, where its sensible heat
is employed in the evaporation of the heat pipe fluid, usually one of the
Freon groups of refrigerants. Incoming cold air is blown through a
completely separate channel and is heated up by the condensation of
this evaporated fluid. In buildings with air conditioning, cold air is recovered in the same way by pre-cooling incoming warm external air in
summer.

Considerable savings can be made in both cases in both installation costs and running costs of heating and cooling equipment.

c Indoor swimming pools Considerable ventilation is required to reduce the moisture content of air. Normally this would be accompanied by appreciable sensible heat losses. Heat pipe heat recovery systems prevent these heat losses.

Industrial heat recovery
Because flow of the source heat medium is completely separate from the supply air flow, heat pipes can be used to recover heat from highly corrosive, poisonous and otherwise objectionable media in industry. Care must be taken in all cases to reduce corrosion and scaling of the evaporator side, and to see to it that adequate means exist to drain off harmful condensate liquids.

As there is a large range of heat pipe fluids suitable for use in industrial equipment, it is possible to operate within any desired temperature range. Equally, commercial equipment exists to deal with volume flows ranging from 250 l/s up to over 40 m^3/s. The only maintenance normally required is cleaning, and even this can be reduced by adequate filtering of the inlet and exhaust air flows.

Q-dot Corporation give the following list of industrial processes from which heat can be abstracted in order to pre-heat air used for space heating in other areas.

Paint drying ovens	Curing ovens
Spray dryers	Forging areas
Boilers	Rubber vulcanizing units
Textile ovens	Plating processes
Desiccant dehumidifiers	Paint spray booths
Brick kilns	Foundries
Paper dryers	Baking ovens
Heat treatment areas	Timber dryers
Reverbatory furnaces	Bleaching ovens
Vinyl ovens	Waste steam exhaust
Casting plant	Grinding areas

Process to process heat transfer
Apart from thermal recovery units, heat pipes are also used in industry for more direct purposes.

Indirect heating and cooling systems
In many industrial processes it is necessary to provide heating or cooling in precise positions at which it is not feasible to place any other heating or cooling devices except heat pipes. As heat pipes can use a variety of fluids, they are capable of universal application. The following are some examples.

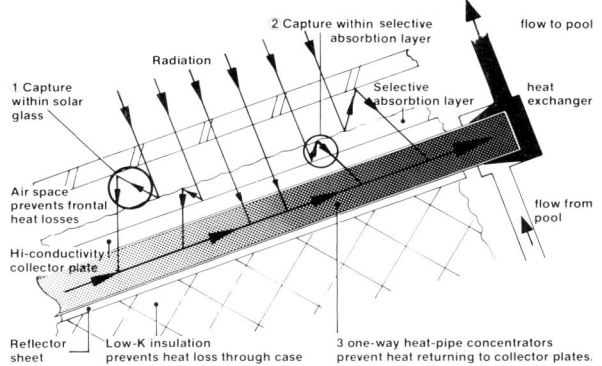

2 Capture within selective
absorbtion layer

Radiation

flow to pool

1 Capture
within solar
glass

Selective
absorbtion layer

heat
exchanger

Air space
prevents frontal
heat losses

flow from
pool

Hi-conductivity
collector plate

Reflector
sheet

Low-K insulation
prevents heat loss through case

3 one-way heat-pipe concentrators
prevent heat returning to collector plates.

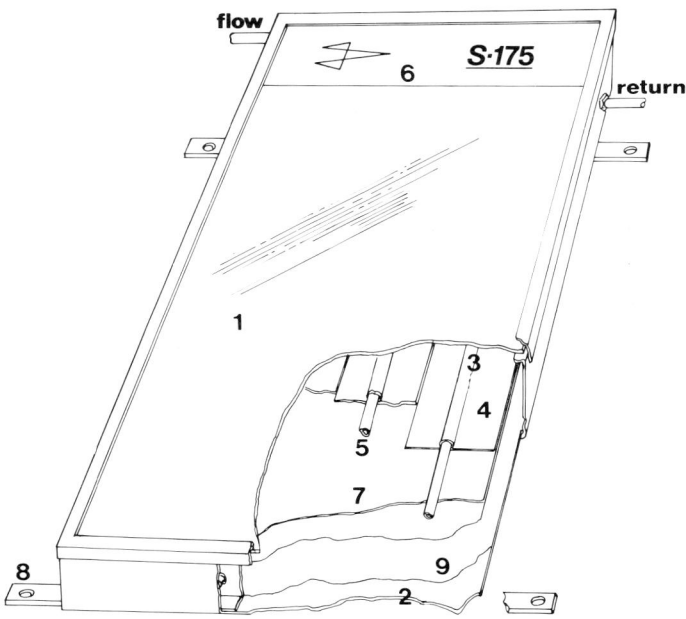

flow

S·175

6

return

1

3

4

5

7

8

9

2

Specification:
Overall sizes: 1.0 x 1.75 x 0.1 metres
(40 x 72 x 4ins)
Mounting centres 1.22m. (48 ins) x 1.45m (57.2 ins)
Area 1.75 m² (20ft²)
Dry weight 42 kg. (92 lbs)
Connections: 10mm pipe

9.8 Heat pipe system used with solar collector to heat swimming pool water.
Courtesy Redpoint Associates.
1 Glass panel
2 Metal casing
3 Heat pipe flow
4 High conductivity collector plate
5 Heat pipe return
6 Heat exchanger system
7 Reflector sheet
8 Fixing lugs
9 Thermal insulation layer

a Cooling electronic components

In complex electronic devices heat often has to be liberated which could cause considerable damage if not dissipated. Heat pipe systems, usually of the plate variety, carry this excess heat from the individual components to a heat sink which is then cooled down by a refrigeration circuit. Such systems are very common indeed and constitute possibly one of the most promising future uses for heat pipes. The main components cooled in this way are transistors, integrated circuit packages and transformers.

b Supply of heat or cold to moulding machines

In many metal or plastic casting processes (eg injection moulding) it is necessary to supply heat to fairly inaccessible parts of the system in order to keep the material fluid. Heat pipes are convenient systems to do this. Equally, in other kinds of casting processes, the liquid metal or plastic is being shaped and must be cooled down rapidly while in the mould. In such cases heat pipes are the most convenient system to carry out such a task.

c Permafrost maintenance

The Trans-Alaska oil pipeline is laid in soil in a permanently frozen condition, known as permafrost. If at any time this permafrost should melt, enormous damage would be caused to the pipeline. To prevent this happening 100 000 heat pipes costing $13 million were placed against the pipeline to enable the soil to be kept frozen even if there should be a thaw at any time.

d Improvement in running of engines

The stirling engine, an engine in which the gas working fluid is heated externally by coal, refuse and numerous other fairly unpromising fuels, is once more being studied after being first invented back in 1816 by a Scottish clergyman. It is necessary to transport heat from a single heat source to the various cylinders of the unit and this has now been found to be achieved most easily by a heat pipe. In the case of normal petrol

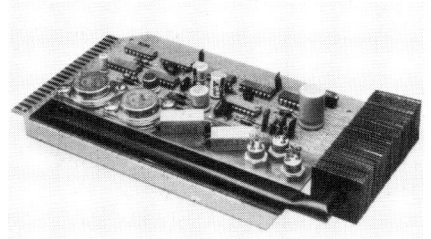
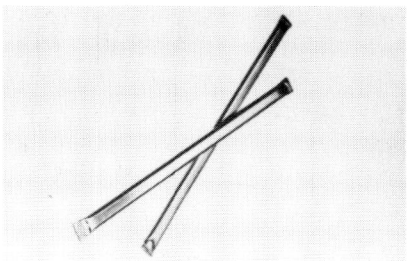

9.9 Heat pipe used to cool electronic components. Courtesy Noren Products
9.10 Mini-heat pipe used for electronic cooling

engines experiments are now being made using heat pipes to transfer exhaust heat to the inlet manifold, to obtain better running.

Numerous other fields of use for heat pipes have been suggested by workers in the field, P. Dunn and D. A. Reay, who list the following additional uses:

Aircraft brake cooling
Gas turbine regenerators
Furnace muffle tubes
Helium target cooling
Cryoprobes for surgery

Solar collectors
Calorimeters
Griddle plates and skewers for cooking
Motor cycle oil cooling
Transformer cooling

No doubt other uses will also be found for this device which has already established itself as an extremely useful tool in its very short period of existence.

SELECTION CHART: HEAT PIPE SYSTEMS

Type of equipment	Advantages	Disadvantages	Appropriate applications
Multiple tube type capillary heat pipes	Comparatively simple construction Heat exchange value 200 times as high as with conventional heat recuperator	Can only be operated horizontally	Most common of all heat recovery system for two gas streams which run side-by-side
Gravity induced fluid flow heat pipes	Cheap because a comparatively small wick is used	Condensing section must be placed well above evaporator section	Widely used where the criterion that the condensing section has to be above evaporator section does not present any problem
Osmotic flow heat pipes	Condenser can be placed well below evaporator level	Low flow rates and low capacity per unit size	Where it is necessary to have condenser below evaporator
Electro-osmotic heat pipes	Ability to pump fluids a long way against gravity	Necessary to use fluids with high dielectric properties High cost	Where there are appreciable distances between condenser and evaporator
Inversion thermosiphon	Evaporator can be placed above condenser	Auxiliary heater needed	Special applications

SELECTION CHART: HEAT PIPE SYSTEMS

Type of equipment	Advantages	Disadvantages	Appropriate applications
Heat plates	Simple design Good heat transfer	Can be used horizontally only	Widely used for general purposes
Flexible heat pipes	Can be used when either evaporator or condenser vibrates	Expensive; limited life	Useful under specific circumstances
Rotating heat pipes	Augment capillary forces by centrifugal force	Limited application	Cooling rotors in electric motors and generators

References and further reading

1 P. D. Dunn and D. A. Reay, *Heat pipes*, Pergamon Press, Oxford, 1976.

2 V. A. Eustace and A. Wright, 'Energy transfer using heat pumps and heat pipes', *Engineering*, September 1980, pp 978–983.

3 D. P. Deyoe, 'Heat recovery—How can the heat pipe help?', *ASHRAE Journal*, April 1973, p 35–38.

D. A. Reay, 'Heat pipe heat exchangers—current status and development potential', *Heating and ventilating engineer*, December 1978/January 1979, pp 8–14.

5 A. Bedrossian and Y. Lee, 'The characteristics of heat exchangers using heat pipes or thermosiphons', *International Jnl. Heat and Mass Transfer*, April 1978, pp 40–44.

6 D. C. Lu and K. T. Feldman, 'Cost-effective study of heat pipe heat exchangers', *Proc ASME Winter annual meeting ASME paper 77 WA/HT-5*, 1977.

7 M. A. Ruch and G. M. Grover, 'Heat pipe thermal recovery unit applications', *Proc II Heatpipe Conference, Bologna*, 1976, ESA report SP112 Vol 1.

8 D. A. Reay, *Directory of heat recover equipment*, Spon, London, 1979.

9 US patent No. 2379548 (The heat pipe patent by R. S. Gaugler), 21 December 1942.

10 US patent No. 3229759 (The heat pipe patent by G. M. Grover), 2 December 1963.

SELECTED COMPANIES INVOLVED IN MANUFACTURE OF PRODUCTS

UK and Europe

Advanced Services Ltd, Unit 2, Hamm Moor Lane, Weybridge Trading Estate, Weybridge, Surrey.

Airdale Air Conditioning Ltd, Clayton Wood Rise, West Park, Leeds LS16 6RF.

Andrews Industrial Equipment Ltd, Dudley Road, Wolverhampton WV2 3DB.

Auchard Development Ltd, Old Road, Southam, Leamington Spa, Warks CV33 0HP.

F. H. Biddle Ltd, Newtown Road, Nuneaton, Warks CV11 4HP.

Clipper Air Handling Units Ltd, Raans Road, Amersham, Bucks HP6 6HY.

Conservatherm Ltd, Hereford House, Station Road, Billinghurst, West Sussex RH14 9SE.

Curwen and Newbery Division KDG Industries Ltd, Fleming Way, Crawley, West Sussex RH10 2QE.

Delchi SpA, Via R. Sanzio 9, 20058 Villasanta, Italy.

Gaylord Industries Ltd, Elizabeth Works, Trinity Trading Estate, Sittingbourne, Kent ME10 2PD.

International Research and Development Ltd, Fossway, Newcastle-upon-Tyne NE6 2YD.

Kiloheat Ltd, Vestry Estate, Sevenoaks, Kent TN14 5EL.

Laidlaw, Drew and Co. Ltd, Sighthill Industrial Estate, Edinburgh EH11 4HG.

Lindsay Engineering Group Ltd, Unit D, Riverside Industrial Estate, Riverside Way, Dartford, Kent.

Matthews and Yates Ltd, Cyclone Works, Swinton, Manchester M27 2AP.

McQuay Perfex Ltd, Curfew Yard, Thames Street, Windsor, Berks.

OY Nokia AB, PO Box 44, 01511 Vantaa 51, Finland.

Paques (UK) Ltd, Shire Hill Industrial Estate, Saffron Walden, Essex CB11 3AQ.

S&P Coil Products Ltd, Evington Valley Road, Leicester LE5 5LU.

Steefane Ltd, Reed House, Bedford Road, Great Barford, Bedford.

Stone Boilers Ltd, Gatwick Road, Crawley, West Sussex R10 2RN.

Temperature Ltd, Newport Road, Sandown, Isle of Wight.

tt Coil ApS, Svanningeved 2, 9220 Aalborg Øst, Denmark.

Walker Air Conditioning Ltd, 30 Sloane Street, London SW1X 9NJ.

Westwarm Ltd, Unit 6, Hither Green, Clevedon, Avon BS21 6XT.

United States

Coil Co. International, 125 S. Front Street, Colwyn, Penna 19023, USA.

Dravo Corporation, PO Box 9305, Pittsburgh, PA 15225.

Dumont Industries, Main Street, PO Box 148, Monmouth, ME 04259.

Energy and Environmental Engineering Inc., 13310 Industrial Park Boulevard, Minneapolis, MN 55441.

Energy Conservation Unlimited Inc., 311 E. Georgia Avenue, PO Box 585, Longwood, FL 32750.

Friedrich Air Cond. and Refn. Co., 4200 N. Pan Am. PO Box 1540, San Antonio, TX 78295.

Frigid Coil/Frick Ind., 13711 Freeway Drive, Santa Fe Springs, CA 90607.

General Resource Corporation, 201 S. Third Street, Hopkins, MN 55343.

Harrison Radiator Div. General Motors Corporation, 200 Upper Mountain Road, Lockport, NY 14094.

J. Howden America, Heat Pipe Products Division, 75 Mill Street, Newton, NJ 07860.

Liebert Corporation, 1050 Dearborn Drive, Columbus, OH 43229.

NRG Enterprises Inc., 6605 Walton Way, Tampa, FL 33610.

Pennwait Corp. Isotron Dept., 3 Parkway, Philadelphia, PA 19102.

Q-dot Corporation, 726 Regal Row, Dallas, TX 75247.

Thermal Components Inc., 2760 Gunter Park Drive, W., Montgomery, AL 36193.

Torin Corporation, Kennedy Drive, Torrington, CT 06790.

Appendix A
Conversion factors and symbols

TABLE A.1 MULTIPLE AND SUBMULTIPLE DESIGNATIONS

Multiples and submultiples		Symbol
10^{12}	tera	T
10^3	giga	G
10^9	mega	M
10^3	kilo	k
10^2	hecto	h
10	deca	da
10^{-1}	deci	d
10^{-2}	centi	c
10^{-3}	milli	m
10^{-6}	micro	μ
10^{-9}	nano	n
10^{-12}	pico	p
10^{-15}	femto	f
10^{-18}	atto	a

The following table gives the conversions from imperial and non-SI metric units into the appropriate SI units. In each case the factor indicates the ratio between the imperial or metric non-SI value and the SI value given.

TABLE A.2 SIMPLE CONVERSION FACTORS

Imperial or non-SI metric unit	Factor	SI unit	Symbol
abampere	10	ampere	A
abcoulomb	10	coulomb	C (As)
abfarad	1	gigafarad	GF (As/V)
abhenry	1	nanohenry	nH (Vs/A)
abohm	1	nanoohm	nΩ (V/A)
abvolt	10	nanovolt	nV
acre	4 046.856	metre^2	m^2
angström	100	picometre	pm
atmosphere	1.013 25 (1 bar = 10^5 N/m^2)	bar	bar

Imperial or non-SI metric unit	Factor	SI unit	Symbol
barn (SI permitted unit)	10^{-28}	metre²	m²
barrel	0.158 987	metre³	m³
board foot	2.359 74	decimetre³	dm³
Btu (British thermal unit) IST	1.055 06	kilojoule	kJ
Btu (mean)	1.055 87	kilojoule	kJ
Btu (thermochemical)	1.054 35	kilojoule	kJ
Btu (39°F)	1.059 67	kilojoule	kJ
Btu (60°F)	1.054 68	kilojoule	kJ
bushel (UK)	36.368 7	decimetre³	dm³
bushel (US)	35.239 1	decimetre³	dm³
calorie (IST)	4.186 8	joule	J
calorie (mean)	4.190 0	joule	J
calorie (thermochemical)	4.184	joule	J
calorie (15°C)	4.185 8	joule	J
calorie (20°C)	4.181 9	joule	J
carat (new)	0.2	gramme	g
carat (old)	0.205 3	gramme	g
chain	20.116 8	metre	m
chaldron	1.309 27	metre³	m³
CHU	1.899 108	kilojoule	kJ
circular mil	506.707 5	picometre²	pm²
cord	3.624 56	metre³	m³
cup	236.588 2	millilitre	ml
curie (SI permitted unit)	3.7×10^{10}	disintegrations/ second	Ci
degree (angle)	17.453 29	milliradian	mrad
dyne	10	micronewton	μN
electronvolt	0.160 21	attojoule	aJ
EMU of current, resistance, etc.	same as abampere, abohm, etc.		
ESU of capacitance	1.112 6	picofarad	pF
ESU of current	333.56	picoampere	pA
ESU of potential	299.79	volt	V
ESU of inductance	0.898 76	terahenry	TH
ESU of resistance	0.898 76	teraohm	TΩ
erg	100	nanojoule	nJ
Fahrenheit (degree)	$°C = (°F - 32) \times 5/9$ $K = (°F - 32) \times 5/9 + 273.15$		
faraday (Carbon 12)	96.487	kilocoulomb	kC
faraday (chemical)	96.495 7	kilocoulomb	kC
faraday (physical)	96.521 9	kilocoulomb	kC
fathom	1.828 8	metre	m
fluid drachm (UK)	3.551 63	centimetre³ (millilitre)	cm³

Imperial or non-SI metric unit	Factor	SI unit	Symbol
fluid dram (US)	3.697	centimetre^3	cm^3
fluid minim (UK)	59.193 9	millimetre^3	mm^3
fluid minim (US)	61.609 0	millimetre^3	mm^3
fluid ounce (UK)	28.413 1	centimetre^3	cm^3
fluid ounce (US)	29.573 5	centimetre^3	cm^3
foot	0.304 8	metre	m
foot^2	0.092 903	metre^2	m^2
foot^3	0.028 316 85	metre^3	m^3
gallon (UK)	4.546 087	decimetre^3 (litre)	dm^3
gallon (US dry)	4.404 884	decimetre^3	dm^3
gallon (US liquid)	3.785 412	decimetre^3	dm^3
gamma	1	nanotesla	nT
gauss	100	microtesla	μT
gill (UK)	142.065	centimetre^3	cm^3
gill (US)	118.294	centimetre^3	cm^3
grain	0.064 798 91	gramme	g
hand	101.6	millimetre	mm
hoppus foot	0.036 054	metre^3	m^3
horsepower (boiler)	9.809 5	kilowatt	kW
horsepower (electric)	0.746	kilowatt	kW
horsepower (550 ftlb/s)	0.745 699 9	kilowatt	kW
horsepower (metric)	0.735 499	kilowatt	kW
horsepower (UK)	0.745 7	kilowatt	kW
horsepower (water)	0.746 43	kilowatt	kW
hundredweight (UK)	50.802 35	kilogramme	kg
hundredweight (US)	45.359 24	kilogramme	kg
inch	25.4 (exactly)	millimetre	mm
inch^2	645.16	millimetre^2	mm^2
inch^3	16.387 06	centimetre^3	cm^3
iron	0.53	millimetre	mm
kilogramme force	9.806 65	newton	N
kip	4.448 222	kilonewton	kN
knot	0.514 444 4	metre/second	m/s
lambert	3.183 099	kilocandela/metre^2	kcd/m^2
langley	41.84	kilojoule/metre^2	kJ/m^2
league (international nautical)	5 556.000	metre	m
league (statute)	4 828.032	metre	m
league (UK nautical)	5 559.552	metre	m
lightyear (SI permitted unit)	9 460.55	terametre	Tm
link	0.201 168	metre	m
maxwell	10	nanoweber	nWb
micron	1	micrometre	μm
mil	25.4	micrometre	μm

Imperial or non-SI metric unit	Factor	SI unit	Symbol
mile (nautical UK)	1 853.184	metre	m
mile (nautical US)	1 852.000	metre	m
mile (statute)	1 609.344	metre	m
mile^2	2.589 988	kilometre^2	km^2
minute (angle)	290.888 2	microradian	μrad
oersted	79.577 47	ampere/metre	A/m
ounce force	0.278 013 9	newton	N
ounce mass (avoirdupois)	28.349 52	gramme	g
ounce troy	31.103 48	gramme	g
parsec (SI permitted unit)	3.083 74 $\times$ 10^{16}	metre	m
peck (UK)	9.092 18	decimetre^3	dm^3
peck (US)	8.809 768	decimetre^3	dm^3
pennyweight	1.555 174	gramme	g
perch (masonry)	0.701	metre^3	m^3
Petrograd standard (timber)	4.672 28	metre^3	m^3
phot	10	kilolumen/metre^2	klm/m^2
pica (printer's)	4.217 518	millimetre	mm
pint (UK)	0.568 261	decimetre^3	dm^3
pint (US dry)	0.550 610 5	decimetre^3	dm^3
pint (US liquid)	0.473 176 5	decimetre^3	dm^3
point (printer's)	0.351 459	millimetre	mm
point (silversmith's)	6.4	micrometre	μ
poise (SI permitted unit)	0.1	newton second/metre^2	N s/m^2
poundal	0.138 255	newton	N
pound (force)	4.448 222	newton	N
pound (mass)	0.453 592 4	kilogramme	kg
pound (troy)	0.373 241 7	kilogramme	kg
quart (UK)	1.136 52	decimetre^3	dm^3
quart (US dry)	1.101 221	decimetre^3	dm^3
quart (US liquid)	0.946 353	decimetre^3	dm^3
quarter (UK)	12.700 6	kilogramme	kg
Rankine degree	1.8	degree Kelvin	K
rod	5.029 2	metre	m
roentgen	0.257 976	millicoulomb/kilogramme	mC/kg
second (angle)	4.848 137	microradian	μrad
slug	14.593 9	kilogramme	kg
Statampere, statohm, etc., same as ESU of current, resistance, etc.			
stokes (SI permitted unit)	1	centimetre^2/second	cm^2/s
stone (UK)	6.350 29	kilogramme	kg
tablespoon	14.786 76	centimetre^3	cm^3

Imperial or non-SI metric unit	Factor	SI unit	Symbol
teaspoon	4.928 922	centimetre^3	cm^3
ton (assay)	29.166 67	gramme	g
ton (UK long)	1 016.047	kilogramme	kg
ton (US short)	907.184 7	kilogramme	kg
ton (nuclear)	4.20	gigajoule	GJ
ton (refrigeration)	3.516 85	kilowatt	kW
ton (register)	2.831 685	metre^3	m^3
ton (shipping UK)	1.189 3	metre^3	m^3
ton (shipping US)	1.132 67	metre^3	m^3
ton (displacement)	0.991 1	metre^3	m^3
torr	133.322	newton/metre^2	N/m^2
township (US)	93.239 57	kilometre^2	km^2
unit pole	125.663 7	nanoweber	nWb
yard	0.914 4	metre	m
yard^2	0.836 127 4	metre^2	m^2
yard^3	0.764 554 9	metre^3	m^3

Combined units

It is obviously quite impossible to give all the combinations of units which can be built up from the values in Table A.2. The required combined units can, however, easily be calculated from the simple units.

For example, let us determine the SI value of

$$\frac{\text{pound force}}{\text{inch}}$$

From Table A.2 we can read off that pound force $= 4.448\ 222$ newton and that inch $= 0.025\ 4$ metre.

$$\text{Therefore,} \quad \frac{\text{pound force}}{\text{inch}} = \frac{4.448\ 222}{0.025\ 4}$$

$$= 175.126\ 8 \text{ newton/metre}$$

Many of the more common compound conversions are given in the following tables.

TABLE A.3 MASS PER UNIT LENGTH, AREA AND VOLUME

Imperial or non-SI metric unit	Factor	SI unit	Symbol
ton (UK)/mile	0.631 342	kilogramme/ metre	kg/m
ton (US)/mile	0.563 698	kilogramme/ metre	kg/m

Imperial or non-SI metric unit	Factor	SI unit	Symbol
pound/yard	0.496 055	kilogramme/metre	kg/m
pound/foot	1.488 16	kilogramme/metre	kg/m
pound/inch	17.858	kilogramme/metre	kg/m
ton (UK)/mile^2	392.298	kilogramme/kilometre^2	kg/km^2
ton (US)/mile^2	350.266	kilogramme/kilometre^2	kg/km^2
pound/foot^2	4.882 43	kilogramme/metre^2	kg/m^2
ounce/foot^2	0.305 152	kilogramme/metre^2	kg/m^2
ton (UK)/yard^3	1 328.94	kilogramme/metre^3	kg/m^3
ton (US)/yard^3	1 186.55	kilogramme/metre^3	kg/m^3
pound/foot^3	16.018 5	kilogramme/metre^3	kg/m^3
pound/gallon (UK)	0.099 776	kilogramme/decimetre^3	kg/dm^3
pound/gallon (US)	0.119 826 4	kilogramme/decimetre^3	kg/dm^3
pound/inch^3	27 679.9	kilogramme/metre^3	kg/m^3

TABLE A.4 FORCE PER UNIT LENGTH, MOMENTS AND MOMENTUM

Imperial or non-SI metric unit	Factor	SI unit	Symbol
ton (UK) force/foot	32.690 3	kilonewton/metre	kN/m
ton (US) force/foot	29.187 8	kilonewton/metre	kN/m
pound force/foot	14.593 9	newton/metre	N/m
pound force/inch	175.127	newton/metre	N/m
ton (UK) force foot	3.037 03	kilonewton metre	kNm
ton (US) force foot	2.711 63	kilonewton metre	kNm
pound force foot	1.355 82	newton metre	Nm
pound force inch	0.112 985	newton metre	Nm
pound foot^2	0.042 140	kilogramme metre^2	kgm^2
pound inch^2	2.926 4	kilogramme centimetre^2	kg cm^2
pound foot/second	0.138 255	kilogramme metre/second	kg m/s

TABLE A.5 PRESSURES

Imperial or non-SI metric unit	Factor	SI unit	Symbol
centimetre of mercury (0°C)	1.333 22	kilonewton/metre²	kN/m²
centimetre of water (4°C)	98.063 8	newton/metre²	N/m²
foot of water (39°F)	2.988 98	kilonewton/metre²	kN/m²
inch of mercury (32°F)	3.386 389	kilonewton/metre²	kN/m²
inch of water (39°F)	249.082	newton/metre²	N/m²
kilogramme force/metre²	9.806 65	newton/metre²	N/m²
poundal/foot²	1.488 164	newton/metre²	N/m²
pound force/foot²	47.880 26	newton/metre²	N/m²
pound force/inch²	6.894 757	kilonewton/metre²	kN/m²
ton (UK) force/foot²	107.252	kilonewton/metre²	kN/m²
ton (US) force/foot²	95.760 7	kilonewton/metre²	kN/m²
ton (UK) force/inch²	15.444 3	newton/millimetre²	N/mm² or MN/m²
ton (US) force/inch²	13.789 55	newton/millimetre²	N/mm² or MN/m²

TABLE A.6 HEAT AND HEAT TRANSFER

Imperial or non-SI metric unit	Factor	SI unit	Symbol
Btu/hour	0.293 071	watt	W (J/s)
Btu/pound	2.326	kilojoule/kilogramme	kJ/kg
Btu/foot³	37.258 9	kilojoule/metre³	kJ/m³
therm/UK gallon	23.208 0	gigajoule/metre³	GJ/m³
therm/US gallon	27.871 7	gigajoule/metre³	GJ/m³
Btu/pound °F	4.186 8	kilojoule/kilogramme K	kJ/kg K
Btu/foot³ °F	67.066 1	kilojoule/metre³ K	kJ/m³ K
Btu/foot² hour	3.154 59	watt/metre²	W/m²
Btu/foot² hour °F	5.678 26	watt/metre² deg C	W/m² deg C
kilocalorie/metre² hour °C	1.163	watt/metre² deg C	W/m² deg C
Btu/foot hour °F	1.730 073	watt/metre deg C	W/m deg C
Btu inch/foot² hour °F	0.144 228	watt/metre deg C	W/m deg C
Btu inch/foot² second °F	519.22	watt/metre deg C	W/m deg C
kilocalorie/metre hour °C	1.163	watt/metre deg C	W/m deg C
calorie/centimetre second °C	418.68	watt/metre deg C	W/m deg C

TABLE A.7 LIQUIDS

Imperial or non-SI metric unit	Factor	SI unit	Symbol
grain/100 foot³	0.022 883 5	gramme/metre³	g/m³
ounce/UK gallon	6.236 02	gramme/ decimetre³	g/dm³
ounce/US gallon	7.489 142	gramme/ decimetre³	g/dm³
pound force hour/foot²	0.172 369	meganewton second/metre²	MNs/m²
pound force second/foot²	47.880 3	newton second/ metre²	Ns/m²
poundal second/foot²	1.488 16	newton second/ metre²	Ns/m²
foot²/hour	25.806 4	millimetre²/second	mm²/s
foot²/second	0.092 903	metre²/second	m²/s
poise	0.1	newton second/ metre²	Ns/m²
stoke	1	centimetre²/ second	cm²/s

TABLE A.8 MISCELLANEOUS COMPOUND CONVERSIONS

Imperial or non-SI metric unit	Factor	SI unit	Symbol
gallon/mile (UK)	2.824 81	decimetre³/ kilometre	dm³/km
gallon/mile (US)	2.352 15	decimetre³/ kilometre	dm³/km
mile/gallon (UK)	0.354 006	kilometre/ decimetre³	km/dm³
mile/gallon (US)	0.425 144	kilometre/ decimetre³	km/dm³
ton (UK) mile	1.635 17	tonne kilometre	tonne km
ton (US) mile	1.459 973	tonne kilometre	tonne km
ton (UK) mile/gallon (UK)	359.687	kilogramme kilometre/ decimetre³	kg km/dm³
ton (US) mile/gallon (US)	321.149	kilogramme kilometre/ decimetre³	kg km/dm³
foot pound force/second	1.355 82	watt	W
foot candle	10.763 9	lumen/metre²	lm/m²
kilovolt/inch	39.370	kilovolt/metre	kV/m

Appendix B
Symbols for pipelines and fittings

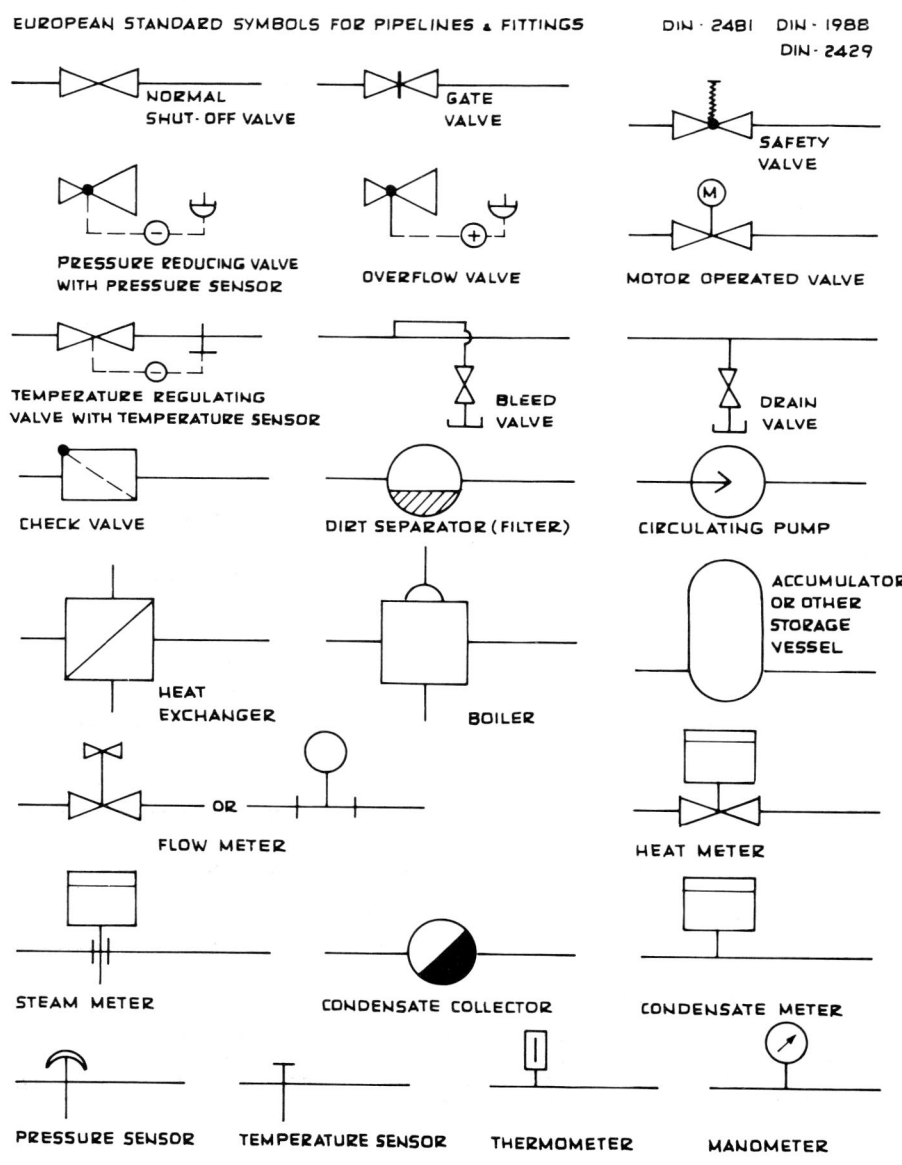

EUROPEAN STANDARD SYMBOLS FOR PIPELINES & FITTINGS   DIN · 2481   DIN · 1988
                                                                        DIN · 2429

NORMAL SHUT- OFF VALVE

GATE VALVE

SAFETY VALVE

PRESSURE REDUCING VALVE WITH PRESSURE SENSOR

OVERFLOW VALVE

MOTOR OPERATED VALVE

TEMPERATURE REGULATING VALVE WITH TEMPERATURE SENSOR

BLEED VALVE

DRAIN VALVE

CHECK VALVE

DIRT SEPARATOR (FILTER)

CIRCULATING PUMP

HEAT EXCHANGER

BOILER

ACCUMULATOR OR OTHER STORAGE VESSEL

OR

FLOW METER

HEAT METER

STEAM METER

CONDENSATE COLLECTOR

CONDENSATE METER

PRESSURE SENSOR

TEMPERATURE SENSOR

THERMOMETER

MANOMETER

AMERICAN STANDARD SYMBOLS FOR PIPELINES & FITTINGS

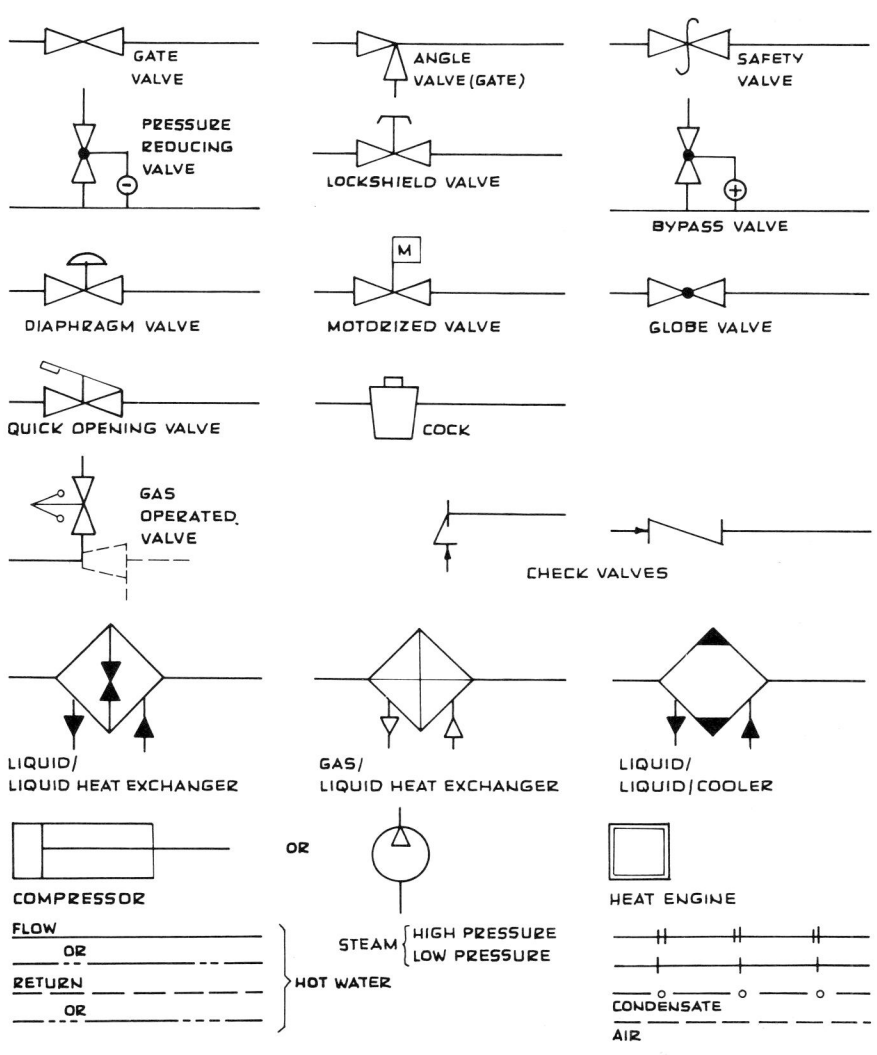

GATE VALVE

ANGLE VALVE (GATE)

SAFETY VALVE

PRESSURE REDUCING VALVE

LOCKSHIELD VALVE

BYPASS VALVE

DIAPHRAGM VALVE

MOTORIZED VALVE

GLOBE VALVE

QUICK OPENING VALVE

COCK

GAS OPERATED VALVE

CHECK VALVES

LIQUID/ LIQUID HEAT EXCHANGER

GAS/ LIQUID HEAT EXCHANGER

LIQUID/ LIQUID/COOLER

COMPRESSOR

OR

HEAT ENGINE

FLOW OR

RETURN OR

STEAM { HIGH PRESSURE / LOW PRESSURE

HOT WATER

CONDENSATE

AIR

NOTE: IF OTHER FLUIDS ARE BEING PUMPED AN UNINTERRUPTED LINE IS USED FOR FLOW & A DASHED LINE FOR RETURN WITH CODE LETTERS IN CENTRE AS UNDER

——— CHF ———
CHILLED WATER FLOW

——— CHR ———
CHILLED WATER RETURN

OTHER ABBREVIATIONS USED ARE:

B - BRINE FLOW
CR- CONDENSER RETURN
A - COMPRESSED AIR
FOF - FUEL OIL FLOW
G - GAS

BR - BRINE RETURN
D - DRAIN
RL - REFRIGERATION LIQUID
FOR - FUEL OIL RETURN
V - VACUUM

C - CONDENSER WATER
H - HUMIDIFICATION LINE
RD - REFRIGERATION DISCHARGE
FOV - FUEL OIL VENT

BRITISH STANDARD SYMBOLS FOR PIPELINES & FITTINGS   BS 2917 - BS 974 - BS M24

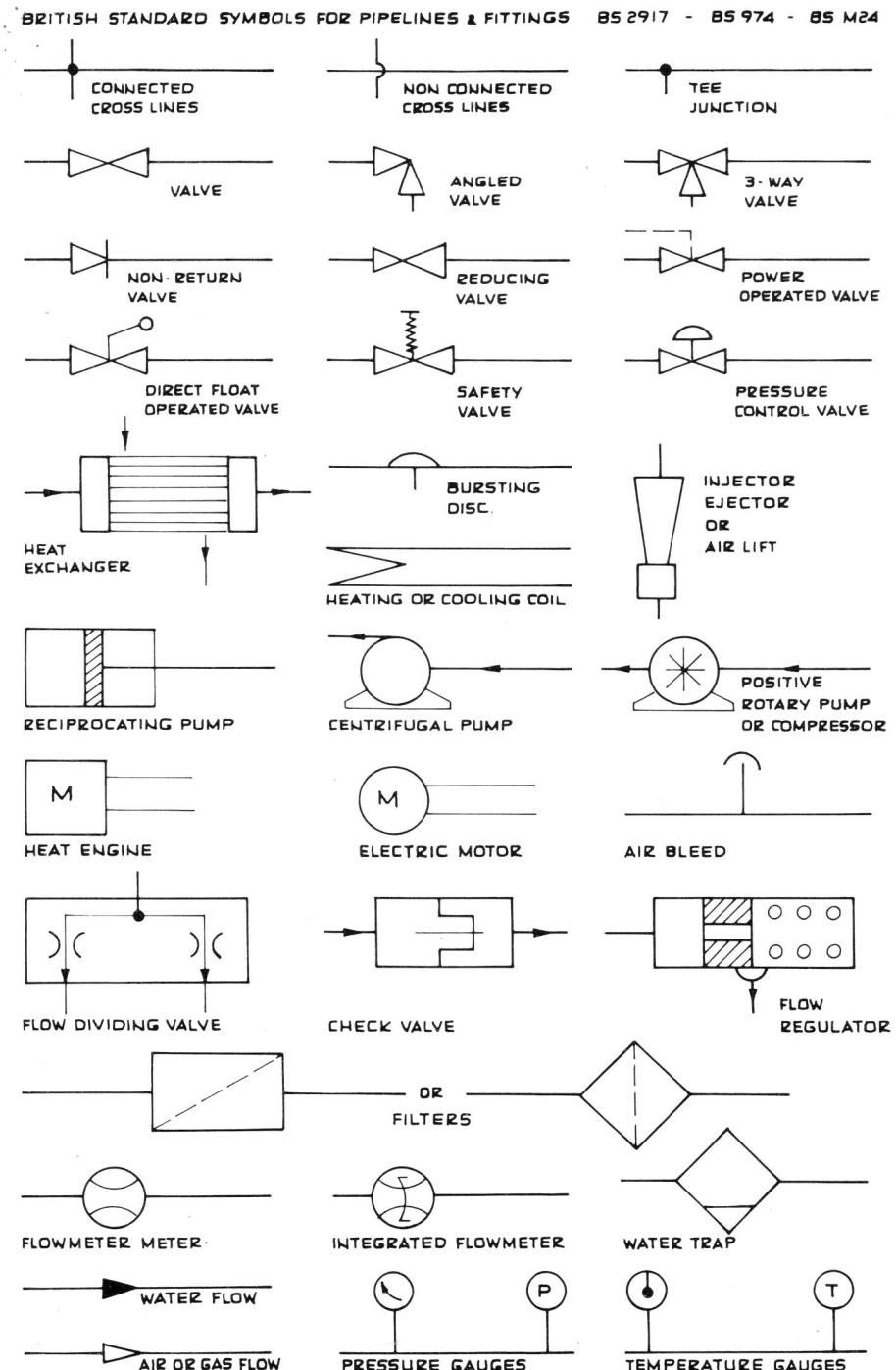

CONNECTED CROSS LINES

NON CONNECTED CROSS LINES

TEE JUNCTION

VALVE

ANGLED VALVE

3 - WAY VALVE

NON - RETURN VALVE

REDUCING VALVE

POWER OPERATED VALVE

DIRECT FLOAT OPERATED VALVE

SAFETY VALVE

PRESSURE CONTROL VALVE

HEAT EXCHANGER

BURSTING DISC

INJECTOR EJECTOR OR AIR LIFT

HEATING OR COOLING COIL

RECIPROCATING PUMP

CENTRIFUGAL PUMP

POSITIVE ROTARY PUMP OR COMPRESSOR

HEAT ENGINE

ELECTRIC MOTOR

AIR BLEED

FLOW DIVIDING VALVE

CHECK VALVE

FLOW REGULATOR

OR FILTERS

FLOWMETER METER

INTEGRATED FLOWMETER

WATER TRAP

WATER FLOW

PRESSURE GAUGES

TEMPERATURE GAUGES

AIR OR GAS FLOW

ENERGY CONSERVATION EQUIPMENT

R. M. E. Diamant
MSc DipChemE MInstE CEng

The Architectural Press: London
Nichols Publishing Company: New York

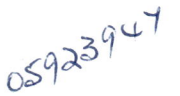

First published in Great Britain 1984 by the
Architectural Press Limited, 9 Queen Anne's Gate,
London SW1H 9BY
Published in the United States of America by
Nichols Publishing Company
New York 1984

British Library Cataloguing in Publication Data
Diamant, R. M. E.
  Energy conservation equipment.
  1. Buildings—Energy conservation
  I. Title
  696   TJ163.5.B84

  ISBN 0–85139–196–6 (British edition)

Library of Congress Cataloging in Publication Data
Diamant, R. M. E. (Rudolph Maximilian Eugen), 1925–
  Energy conservation equipment.

  Includes bibliographical references.
  1. Energy conservation.   2. Power (Mechanics)
  I. Title.
  TJ163.3.D49   1984   621.402′5   83–27014
  ISBN 0–89397–190–1 (United States edition)

Typeset by Phoenix Photosetting, Chatham
Printed in Great Britain by Biddles Ltd, Guildford, Surrey